Computational Social Sciences

Computational Social Sciences

A series of authored and edited monographs that utilize quantitative and computational methods to model, analyze and interpret large-scale social phenomena. Titles within the series contain methods and practices that test and develop theories of complex social processes through bottom-up modeling of social interactions. Of particular interest is the study of the co-evolution of modern communication technology and social behavior and norms, in connection with emerging issues such as trust, risk, security and privacy in novel socio-technical environments.

Computational Social Sciences is explicitly transdisciplinary: quantitative methods from fields such as dynamical systems, artificial intelligence, network theory, agent-based modeling, and statistical mechanics are invoked and combined with state-of-the-art mining and analysis of large data sets to help us understand social agents, their interactions on and offline, and the effect of these interactions at the macro level. Topics include, but are not limited to social networks and media, dynamics of opinions, cultures and conflicts, socio-technical co-evolution and social psychology. Computational Social Sciences will also publish monographs and selected edited contributions from specialized conferences and workshops specifically aimed at communicating new findings to a large transdisciplinary audience. A fundamental goal of the series is to provide a single forum within which commonalities and differences in the workings of this field may be discerned, hence leading to deeper insight and understanding.

More information about this series at http://www.springer.com/series/11784

Sara McComb • Deanna Kennedy

Computational Methods to Examine Team Communication

When and How to Change the Conversation

Springer

Sara McComb
School of Nursing and School
of Industrial Engineering
Purdue University
West Lafayette, IN, USA

Deanna Kennedy
School of Business
University of Washington Bothell
Bothell, WA, USA

ISSN 2509-9574 ISSN 2509-9582 (electronic)
Computational Social Sciences
ISBN 978-3-030-36158-7 ISBN 978-3-030-36159-4 (eBook)
https://doi.org/10.1007/978-3-030-36159-4

This Springer imprint is published by the registered company Springer Nature Switzerland AG
The registered company address is: Gewerbestrasse 11, 6330 Cham, Switzerland

To Abhi and Kip

Acknowledgments

This book is the culmination of my almost 20 years spent studying team communication and cognition. Throughout these years, many people have been instrumental as I developed and conducted this research.

Mike Letsky, who saw something in the first proposal I wrote to the Office of Naval Research, brought me into a community of scholars whose constructive comments helped me become a better researcher. Rick Alterman, Steve Kozlowski, Joan Rentsch, Eduardo Salas, and Norm Warner are a few of the many colleagues who encouraged, supported, and mentored me during those years. To all of them, I say thank you.

My research program would not have been possible without the support and commitment of many students and colleagues from the University of Massachusetts, Texas A&M University, and Purdue University. They have contributed in a variety of ways, and for that, I am deeply thankful. Four people, in particular, played critical roles in making this book a reality. Ralitza Vozdolska was with me at the beginning; she helped set up the lab, design the task, collect data, and so much more. Veronica Muñoz-Najar helped code all the data, which was an enormous task. Zach Hass willingly shared his statistical knowledge when I needed to be sure we were heading in the right direction. Nathan Mueller prepared and organized many of the graphics included in this book. I am indebted to them.

I could not have written this book without my coauthor, Deanna Kennedy. She has contributed in countless ways to this research as she evolved from graduate student to valued collaborator. She deserves many thanks.

Scott Poole encouraged us to write this book. About a year ago, I asked him if we were crazy to consider heading down this path, and he replied, "of course you are... everyone who writes a book is crazy, but it is good crazy." Now that we are on the other side, I couldn't agree more and am truly grateful for his encouragement.

Finally, I would like to thank my husband, Abhi Deshmukh. His unwavering belief in me has made all the difference.

Sara McComb

I would like to thank Sara McComb for keeping us organized and on track to bring this project to fruition. I would also like to extend my thanks to our group and team research colleagues who were audience members, reviewers, and supporters of our methodologies and study progress over the years. Finally, I wish to thank my husband, Kip Kennedy, for his support and encouragement throughout this and many other research projects.

Deanna Kennedy

Portions of this research were supported in part by the National Science Foundation and the Office of Naval Research.

Contents

About the Authors

Sara McComb is a professor at Purdue University with a joint appointment in Nursing and Industrial Engineering. She earned her PhD at Purdue University in the School of Industrial Engineering. Prior to returning to Purdue, she was a member of the faculties at the University of Massachusetts Amherst and Texas A&M University. Her research focus is on team communication and cognition. In particular, she is interested in *when* specific activities may benefit team processes and performance, which extends beyond other researchers' foci on *what* and/or *how* activities transpire. From a methodological perspective, she is one of the very few scholars who has advanced the field by applying computational modeling and simulation to the examination of team behavior.

Deanna Kennedy is an associate professor in the School of Business at the University of Washington Bothell. She received her PhD in Management Science from the Isenberg School of Management, University of Massachusetts Amherst. Her research interests are in using communication and other processes to advance group and team effectiveness for those teams engaged in supply chain, operations, and project management. Her work identified team process shift points within communication where there is a potential for intervention strategies to help teams work toward an optimal benchmark. She is a proponent of computational methods including simulation and optimization that can uncover new and productive areas for inquiry.

List of Abbreviations

AIC	Akaike Information Criterion
ANOVA	Analysis of Variance
ARIMA	Auto Regressive Integrated Moving Average
CAR	Consultation Analysis Record
CoPrA	Collaboration Process Analysis
DDV	Dynamic Dummy Variables
DET	Determinism
ENTR	Entropy
LSA	Latent Semantic Analysis
M	Mean
MANCOVA	Analysis of Covariance
Q1	Quartile 1
Q2	Quartile 2
Q3	Quartile 3
Q4	Quartile 4
RQ	Research Question
RQA	Recurrence Quantification Analysis
RR	Recurrence Rate
SD	Standard Deviation
T1	Performance Episode 1
T2	Performance Episode 2
UAV	Unmanned Aerial Vehicle

List of Figures

List of Tables

Chapter 1
Team Communication in Theory and Practice

Why Write a Book About Longitudinal Team Communication?

Teams are ubiquitous in organizations. They are useful because as temporary systems comprised of members attending to time-limited projects (Goodman and Goodman, 1976), they can be reconfigured as needed to address short- and long-term challenges. Teams are often constructed of members representing different disciplinary perspectives who must collaborate interdependently to transform inputs into outputs (Ilgen et al. 2005; Lepine et al. 2008; Mathieu et al. 2008), thereby working to achieve predetermined goals established by the organization. The purpose of establishing these goals is to change the current state to some future, more profitable state. These changes could result in the design and development of new products, improved processes, market analyses, or any number of outcomes that may require the combined perspectives of an interdisciplinary team of people.

Researchers have focused their efforts on studying groups and teams for many years. Yet, much of this research represents team processes statically, even though teams function dynamically. Team scholars acknowledge the lack of conceptual and analytical frameworks available to guide examinations of continuous team behavior (Leenders et al. 2016) and limited empirical evidence from studies to explain temporal phenomena (Cronin et al. 2011; Ilgen et al. 2005; Kozlowski et al. 2016). Simultaneously, scholars call for more longitudinal research focused on team processes (Kozlowski et al. 2013; Mathieu et al. 2008; Sonnentag 2012). Indeed, in a recent paper considering team process dynamics, Kozlowski (2015) states that "empirical examination of team dynamics is on the research frontier" (p. 270).

To address these calls for research, several journals have hosted special issues dedicated to longitudinal investigations of team behavior including *Human Factors*' issue on collaboration, coordination, and adaptation in complex sociotechnical settings in 2010 (volume 52, issue 2), *Group Dynamics*' issue on statistical methods in

S. McComb, D. Kennedy, *Computational Methods to Examine Team Communication*, Computational Social Sciences,
https://doi.org/10.1007/978-3-030-36159-4_1

group psychology and group psychotherapy in 2016 (volume 20, issue 3), *Small Group Research*'s issues on group learning (volume 48, issue 2) and interdisciplinary insights into group and team dynamics (volume 48, issue 5) in 2017, and the *Journal of Organizational Behavior*'s issue advocating that "the time has come to study dynamics at work" (volume 39) in 2018. These foci on dynamic investigations of human behavior, in general, and team behavior, more specifically, demonstrate the editorial boards' acknowledgements of the need for more research in this area and commitments to publishing it.

Nevertheless, the research examining dynamic team behavior is insufficient. Team scholars have concurred for many years that different processes may be relevant at different times (Hackman and Wageman 2005; Marks and colleagues' 2001) and the way activities unfold early in collaboration may have bearing on later team processes and, ultimately, on team performance (Cohen and Bailey 1997; Ericksen and Dyer 2004; Gersick 1988). Yet, the nuances that cause these phenomena are not well understood. Therefore, more research examining team processes over time is warranted.

Team communication is a particularly significant team process because it is the vehicle through which the majority of collaboration is accomplished. Moreover, what teams discuss provides insights into the processes being enacted by the members (Futoran et al. 1989; von Cranach et al. 1986). For example, team members who state, "I will look into …" or "How should we approach …? " or "Are we responsible for …? " offer insights about processes required to take action, devise a strategy, and assimilate the project's scope, respectively. Missing from our understanding, however, is evidence about when and how often these types of processes are necessary. Our aim in this book is to address this gap by examining the longitudinal team communication exchanged among team members collaborating under different working conditions.

Longitudinal Team Communication Research

Longitudinal team communication appears periodically in the research literature. Much of that research, however, provides limited insight into how that communication progresses over time. Indeed, most researchers aggregate communication into discrete counts, typically one number, that represents the frequency of communication. In healthcare, these counts are often used to examine communication before and after an intervention (e.g., Gittinger et al. 2017; Hua et al. 2012; Mancuso et al. 2016; Mundt and Zakletskaia 2018). Other healthcare researchers have coded communication content, but then aggregated frequency counts by content (e.g., Barth et al. 2015a, b; Michelson et al. 2017; Santos et al. 2012) or by type of communication such as instructions or questions (e.g., Xiao et al., 2003). In educational settings, including both laboratory and classroom-based studies, researchers have aggregated various aspects of communication including word counts (e.g., Caldwell and Everhart 1998; Kanawattanachi and Yoo 2007; McKendrick et al. 2014), message frequency (e.g., Espinosa et al. 2015; Kanawattanachi and Yoo 2007; McKendrick et al. 2014; Swigger et al. 2012), frequency counts by content area (e.g., Cooke et al. 2007; McKendrick et al. 2014; Priest et al. 2004), and turn taking (e.g., Espinosa et al. 2015). Even though some of these studies examine frequency counts over two or more time periods (e.g., Barth et al. 2015a, b; Hua et al. 2012; Swigger et al. 2012) and the results contribute to our extant understanding of team communication behavior, the findings do not provide a nuanced view of communication over time.

Another approach to examining longitudinal team communication behavior has been via questionnaires that elicit self-report data about perceived communication frequency. These studies have assessed a variety of topics including perceived communication before and after interventions in healthcare settings (e.g., Bunger and Lengnick-Hall 2018; Pandhi et al. 2018; Staveski et al. 2017; Wittenberg et al. 2016), changes in perceived communication frequency over time among members of new product development teams (Griffin and Hauser 1992), the relationship between physical distance and perceived communication frequency (Van den Bulte and Moenaert 1998), and the impact of turnover and tenure on perceived team communication (Hirst 2009). As with frequency counts, these studies provide insight into important aspects of team behavior, but do not capture the actual communication taking place.

Studies that examine actual communication content over time are limited. Even many these studies, however, do not delve into the dynamic nuances of communication over time. For instance, researchers have focused on situation analyses within healthcare to identify opportunities for improving team communication and minimizing the occurrences of adverse events attributable to communication breakdowns during activities such as transitions of care (Roshotte et al. 2016) or among members of resuscitation teams working under time pressure (Calder et al. 2017).

Other researchers use actual communication content over time to ascertain the alignment between entities, as two studies demonstrate. First, Ekker (2016) examined terminology alignment across different agencies involved in emergency response and compared the alignment in teams from Norway and Sweden. They coded specific emergency response terminology spoken during emergency management training and conducted cluster analysis. The findings suggest that emergency responders in Norway have aligned their terminology across agencies, whereas those in Sweden have commonality across groups of agencies (i.e., police and fire use similar terminology as do emergency operators and power companies, but the terminology does not align across the two groups). Second, Entin and colleagues (2003) employed survival analysis to examine time between when various categories of communication were exchanged. They found that when team organizational structures aligned with the requirements of the task, the communication rates were different than when teams experienced misalignment.

Neither the situation analyses nor the alignment studies provide holistic, generalizable findings about how communication content patterns emerge and/or evolve over time. Studies that begin to fill this gap are reviewed in Chap. 2. The researchers conducting these studies have employed a variety of methods to provide evidence about the impact of, for example, timing, work requirements, collaboration media, and team member characteristics on communication, as well as the influence of team communication on performance and other outcomes. Within this body of work, several research groups have made multiple contributions to this area of inquiry including Cooke and Gorman (e.g., Gorman and Cooke 2011; Gorman et al. 2012a, b), Miller (e.g., Miller et al. 2009; 2010), and Kennedy and McComb (e.g., Kennedy and McComb 2014; McComb et al. 2010). But, only one research team has systematically targeted one aspect of team activity: Kauffeld and Lehmann-Willenbrock have focused on team meetings. The contributions from their considerable efforts range from indications about what order topics should be discussed (Kauffeld and Meyers 2009) to how to inhibit dysfunctional behaviors through the types of communication exchanged (Lehmann-Willenbrock and Allen 2014) or the disposition of team members (Lehmann-Willenbrock et al. 2016) to how team leaders can guide communication (Lehmann-Willenbrock et al. 2015) to the influence of communication patterns on team-level outcomes (e.g., Lehmann-Willenbrock and Allen 2014).

The research we present in this book provides another example of a holistic investigation into the dynamics of longitudinal team communication. Rather than focusing on a particular team activity (e.g., team meetings), our research focuses on the communication required to complete an assigned task from beginning to end and then repeat the process for a slightly modified task during a second performance episode. We employ a variety of analytical methods to garner multiple perspectives about the way in which communication unfolds over time and if the patterns of communication impact team performance. This approach, like the work of Kauffeld and Lehmann-Willenbrock, makes a robust contribution to team science, in general, and the team communication literature, in particular, through our systematic examination of longitudinal team communication.

What Theoretical Frameworks Help Explain Longitudinal Team Communication?

The theoretical underpinnings of our research are Marks and colleagues' (2001) recurring phase model of team processes and Bartunek and Woodman's (2015) temporal dimensions of organizational change.

Recurring Phase Model of Team Processes

Marks and colleagues' (2001) conceptualize team processes as a series of performance episodes, where performance episodes represent discrete periods within the team's lifecycle that have an identifiable beginning and end. Performance episodes are comprised of recurring phases that occur sequentially and simultaneously as teams work to achieve their goals. During these phases teams engage in transition, action, and interpersonal processes. Specifically, team activities cycle between transition and action processes, with interpersonal processes interspersed throughout that facilitate interpersonal relationships among the team members (e.g., managing conflict, motivating). For the purposes of this research, we focus on the transition and action processes that (1) are observable through team communication messages and (2) occur as teams transit through two performance episodes.

The transition processes are comprised of planning and evaluative activities that set direction for the team's accomplishment of its goals. The specific transition processes are:

- Mission Analysis: interpreting and evaluating the team's assigned mission; activities may include establishing the main tasks required of the team, surveying current environmental conditions vis-à-vis the mission, and ascertaining the resources needs.
- Goal Specification: identifying and prioritizing the goals required for the team to accomplish its mission.
- Strategy Formulation and Planning: developing the approach the team will use, including optional courses of action, to accomplish its mission.

The action processes are comprised of the actual work required to achieve the team's mission. The specific action processes are:

- Monitoring Progress Toward Goals: assessing team activities as they relate to the specified goals and providing feedback to the team about its progress.
- Systems Monitoring: paying attention to team resources and environmental conditions that may impact the team's ability to complete its mission.
- Team Monitoring and Backup Responses: supporting team members as they work to complete the mission, which may include providing feedback, assisting with task completion, and/or assuming responsibility for specific tasks.
- Coordination: organizing the team members' interdependent activities.

This recurring phase model is prevalent within the team literature. It serves, for example, as a cornerstone of conceptual frameworks (e.g., Brown et al. 2018; Reynolds and Salas 2016), provides a theoretical foundation for survey instrument development (e.g., Littlepage et al. 2016), informs coding frameworks for interviews (e.g., Diaz Granados et al. 2018; Marks et al. 2005) and videos (e.g., Fernandez et al. 2013; Marks et al. 2005), is used to explain results in qualitative studies (e.g., Fleştea et al. 2017; Morales and Martin 2018), and guides the design of training and evaluation activities (e.g., Brooks-Buza et al. 2011). Researchers apply the model in various ways by including the team processes used within each phase (e.g., Benzer et al. 2016; Brooks-Buza et al. 2011), examining activities at the transition, action, and interpersonal level (e.g., Wybo 2013), aggregating all included dimensions to represent team processes (e.g., Fernandez et al. 2013; Marks et al. 2005), incorporating only specific parts of the model such as coaching and backup behaviors (e.g., McEwan et al. 2018, Welp et al. 2018) or only transition and action processes (e.g., Fernandez et al. 2013; Kennedy and McComb 2014; Marks et al. 2005; Wybo 2013).

Much of the research is cross-sectional (e.g., Benzer et al. 2016; Diaz Granados et al. 2018; Littlepage et al. 2016) or provides a holistic assessment of team processes over the team's life cycle (e.g., Brooks-Buza et al. 2011; Marks et al. 2005). Very few researchers have tested the episodic nature of the model. Those who have tested it have done so using a variety of approaches, as demonstrated by the following examples. One approach has been to divide the data into naturally occurring performance episodes that were identified inductively through descriptions of the team activities. For example, Fleştea and colleagues (2017) created a chronology of search and rescue activities from interview data and identified the performance episodes, and corresponding interactions among team members, occurring over time. Their results suggest that the efficacy of transition processes has an impact on the team's ability to effectively execute action processes. Alternatively, Wybo's (2013) examination of teams in Paris, Panama, and Haiti orchestrating the Haiti earthquake disaster relief effort, identified regularly scheduled meetings at the start of each team's workday as the transition processes necessary to ensure a common understanding of the current state and prioritize the day's activities. His results suggest that the dynamic nature of the crisis may not result in distinct cycles and not all players may be involved in all phases focused on transition processes.

Another approach has been to divide the data into multiple time periods; researchers then compare and contrast the activities occurring in each period. For instance, van Ruijven and colleagues (2015) examined the network communication patterns among team members and characterized the activities occurring in the phases focused on transition and action processes as centralized or decentralized coordination. Their findings suggest that a more fine-grained delineation than transition and action may be needed to understand longitudinal team processes and differences may occur over time as they observed less coordination (i.e., more action) during the middle time period. DeChurch and Haas (2008) designed an experimental task with two distinct performance episodes and crafted questionnaire items to assess two transition processes (i.e., deliberate and contingency planning) and two action processes (i.e., reactive strategy adjustment and coordination). Their findings suggest that (1) transition processes facilitate action processes, particularly in the earlier performance episode and (2) reactive strategy adjustment may be the most influential team process examined and it predicts coordination and performance.

In two final examples the researchers used coding schemes that incorporated aspects of Marks and colleagues' (2001) model and conducted longitudinal analyses to ascertain how team processes evolve over time. First, Havakhor and Sabherwal (2018) examined the emergence of transactive memory systems. Their coding scheme incorporated many of the team processes from all three phases of Marks and colleagues' (2001) model and other dimensions relevant to their interest in transactive memory systems. Their primary finding related to the recurring phase model of team performance is that positive interpersonal processes, in conjunction with high levels of interaction among team members, facilitate more constructive monitoring (i.e., task versus team member focused) at later stages of the team's lifecycle.

Second, Kennedy and McComb (2014) examined the timing of when teams shift between transition and action processes by using a coding scheme that included the identified team member messages as action or one of the specific transition processes (e.g., mission analysis). They identified when teams shifted from their first substantial discussion of each transition process and communicated their first instance of action. The interrelationships among these points and the relationships between these points and team performance were examined. Virtual teams were generated using computational modeling and simulation procedures and interventions were tested to determine the impact of initiating conversations about mission analysis or strategy formulation at the onset of team activity and/or delaying the start of action. The findings suggest that delaying action until the team's midpoint resulted in the best team performance, regardless of where team conversations start.

Temporal Dimensions of Organizational Change

Bartunek and Woodman (2015) advance a temporal model of change that fundamentally shifts the organizational development paradigm employed to examine planned organizational change. Historically, Lewin's (1947) unfreeze-change-refreeze diagnostic model of organizational change dominated this field. Researchers, however, have been shifting away from this linear, episodic, prescriptive paradigm. Bartunek and Woodman (2015) use these shifts within the field as a foundation for constructing a more temporal representation of the organizational change process. Ultimately, they recommend taking an emergent perspective on organizational change that encompasses the temporal processes required to achieve it. One key component of this perspective is the use of dialog to understand the current state and implement any planned interventions.

Such dialog does not occur linearly. Correspondingly, static research approaches may not fully capture the intricate mechanisms that underlie the change process. Therefore, Bartunek and Woodman (2015) advance five temporal dimensions that describe how processes may evolve over time, which may be useful in both facilitating and researching organizational change. These dimensions are an effort to capture the dynamic, non-linear manner in which change manifests in organizations. The specific temporal dimensions are:

- Rhythm: repetitive cycles of change
- Timing: when events occur
- Pacing: speed of change
- Polyphony: aligned and overlapping events
- Sequence: the temporal ordering of events

Bartunek and Woodman (2015) advocate using their framework beyond organizational change. For instance, they apply it to Tushman and Romanelli's (1985) conceptualization of organization-level convergence that guides the forward momentum and incremental change experienced within organizations until a time when the company needs to reorient itself and reconverge. They also suggest that "process models almost by definition explore change" (p. 176), but that change resulting from such processes will most likely be emergent versus their paradigmatic focus on planned change. Nevertheless, integrating temporal dimensions into process models will facilitate more in depth examinations. To date, no known published research applies this framework to teams.

Application to Longitudinal Team Communication

The theoretical frameworks forwarded by Marks and colleagues (2001) and by Bartunek and Woodman (2015) are complementary in that they represent what team processes occur and how teams might progress through them, respectively.

These frameworks are relevant for examinations of team communication as it may be a window through which researchers can observe team processes, such as the transition and action processes that comprise Marks and colleagues' (2001) recurring phase model. Indeed, researchers of team communication have recognized for over thirty years that team processes may be regulated and coordinated through the words that members use to communicate with one another (von Cranach et al. 1986). That is, the content of communication is inherently connected to team interaction processes, and may therefore serve as a mechanism for assessing patterns of interactions over time (Futoran et al. 1989).

Marks and colleagues (2001) posit that teams will move through related performance episodes to bring about some change in the current state that progresses them toward their goals. The episodes will be comprised of a series of recurring transition and action processes, interspersed with interpersonal processes. The transition processes occur when the team members work on issues relating to their mission analysis, goal specification, and strategy. Action processes transpire when the team members are actively engaged in executing their task. When new information becomes available, a team may need to iterate among the transition processes before returning to action. Therefore, employing this framework to examine longitudinal team communication offers the opportunity to garner insights about how teams interact to achieve their goals.

Performance episodes, as conceptualized by Marks et al. (2001), are comprised of specific processes. Those processes may vary in a variety of ways, including which processes are relevant at particular points in time, how interrelated they are, how long they are focal, and how repetitively they are enacted. Reframing these potential variations in terms of Bartunek and Woodman's (2015) dimensions, we expect transition and action processes to cycle over time (rhythm), have different frequencies of occurrence over time (timing), be iteratively discussed (pacing), affect one another (polyphony), and differ in how they are interwoven into conversations (sequence).

For our purpose we investigate longitudinal team communication guided by the theoretical contributions of Marks and colleagues' (2001) and Bartunek and

Woodman (2015). Specifically, we have coded communicated messages from team transcripts as one of Marks and colleagues' (2001) transition processes or action, as described in Chap. 3. The coded messages are represented in communication strings, that can be examined to investigate the rhythm, timing, pacing, polyphony, and sequence of team communication.

What Team Types Are Examined in this Book?

As Salas and colleagues' (2018, 595) posit, "all teams are *not* created equal" because "context matters." The influence of many different contextual factors on team communication have been examined including communication media, geographic distribution, hierarchical structure, situational stressors, task type, team familiarity, team size, and training (Tiferes and Bisantz 2018). In our research, we focus on two aspects of context that are relevant for teams in the field and may influence how the members communicate as they collaborate. First, we consider differences associated with the medium teams use to communicate, where teams engage in distributed, synchronous communication via an electronic collaboration space or communicate in person synchronously around a table. Second, we examine differences attributable to the amount of time available to complete the task, with teams being either unrestricted or time pressured.

Distributed Versus In Person Teams

Team members are routinely distributed across various locations as they complete their assigned tasks. Online tools are being developed constantly to support these teams. For example, text-based tools include Slack, HipChat, and Google Hangouts Chat; video conferencing tools include WebEx, GoToMeeting, and Zoom. In our study, teams communicated synchronously either via a text box in an electronic collaboration space or in person around a table. These different media types are anticipated to influence how information is exchanged among the team members.

Media is a relevant consideration when interested in understanding more about how teams communicate because the media may help or hinder the flow of information. Communicated information travels through noise-filled channels from sender to receiver (Shannon and Weaver 1949) and media may foster or impede that travel (Daft and Lengel 1986). Daft and Lengel's (1986) media richness theory conceptualizes how various communication media may impact the flow of information. At one extreme, lean communication media, such as online text-based communication, may limit a team's ability to convey information. In particular, social cues may not transmit well via lean media (Straus and McGrath 1994). Rich media, alternatively, provides many verbal and non-verbal (e.g., intonation, facial expressions, paralinguistic cues) channels through which communication can be transmitted.

The inference is that communication via rich media is easier to construct and interpret. Lean media, however, may also have advantages as it encourages more deliberate and parsimonious exchanges. In our research, teams communicating via text in our distributed communication condition were subjected to using a lean medium. Conversely, teams communicating in person were collaborating through a rich medium.

Many researchers have compared distributed and in person teams. The results, overall, are mixed. For example, researchers have found that distributed teams struggle to make accurate decisions (Thompson and Coovert 2003), and take longer to make decisions but those decisions are more accurate than the decisions made by in person teams (Pridmore and Phillips-Wren 2011). Others found that distributed teams make better decisions but only when members perceived themselves to be open to experiences (Colquitt et al. 2002). Distributed teams, in comparison to in person teams, may also have difficulty developing collective cognition (Andres 2013), engage less frequently over longer periods of time (Hammond et al. 2005), and be less satisfied (Thompson and Coovert 2003). Simultaneously, distributed teams may be at an advantage over in person teams because they generate more ideas (Purvanova et al. 2014), have more active and balanced member participation (Berry 2006), and are less likely to have task conflict evolve into disruptive relationship conflict (Martinez-Moreno et al. 2012). Although these results are equivocal, they do suggest that differences exist between teams interacting via different media.

We are interested in identifying any differences in dynamic longitudinal communication between distributed and in person teams. Only one study identified in our structured literature review (see Chap. 2), compared the communication patterns of distributed and in person teams. McComb and colleagues (2010) found that distributed teams, when compared to teams collaborating in person, discuss certain topics later and address topics more serially. No other dynamic, longitudinal studies of communication were found that directly compared these two types of teams. Our study, therefore, will contribute to this literature through its multi-faceted examination of similarities and differences in how conversations unfold over time between teams using these two distinct communication media.

Unrestricted Versus Time Pressured Teams

Project teams, by their very nature, are time limited. The amount of time available, however, may range from reasonable to constrained. Strategically planned initiatives, such as new product development, often are scheduled to provide teams with the requisite time to successfully complete their tasks. Conversely, teams may be brought together to address emergent situations, such as creating a software update to patch an unanticipated security issue, that may reduce the amount of time available for teams to complete their work. In our study, unrestricted teams were allowed to complete their tasks without any time restrictions and time pressured teams were given an abbreviated amount of time to complete the same tasks.

In general, researchers suggest that time pressure is detrimental to team performance (e.g., Driskell et al. 1999; Ellis 2006; Kelly and McGrath 1985), which may be attributed to the sense of urgency members place on making task progress under time pressure. Essentially, members may disrupt the normal way team processes unfold by rushing procedures and ultimately, creating process losses that limit a team's ability to achieve its potential (Steiner 1972). For example, teams may have (1) difficulty identifying and agreeing upon the team's mission, (2) less thorough analysis of potential alternative approaches, solutions, etc., and (3) limited time for the team to socialize, which may inhibit finding a communion language and appreciating the variations in functional perspective (Berry 2006).

Communication may be affected when teams face time pressure. Evidence suggests that time-pressured teams may reduce the number of topics discussed (Karua and Kelly 2004), limit the amount of information shared (Chong et al. 2005), and spend more time on task-relevant activities than they do on activities tangential to the task (Karau and Kelly 1992). The type of communication exchanged may also shift, with teams in high urgency situations exchanging more instructions and less questions than those in low urgency situations (Xiao et al. 2003). Time pressured teams may adapt to their situation and increase the efficiency of their communication (Pfaff 2012). Interestingly, when combined with media differences, time pressure has been found to be more impactful on in person than distributed teams (Caballer et al. 2005).

Our aim in this research is to identify differences in the patterns of communication over time between teams with unrestricted time and those collaborating under time pressure. We identified several studies through our structured literature review (see Chap. 2) that examined teams in situations that could involve time pressure (e.g., training simulations that include emergency scenarios (Davis et al. 2016; Dunn et al. 2002; Stachowski et al. 2009; van der Haar et al. 2017), or airplane pilots reacting to an inflight emergency (David and Schraagen 2018)). The patterns identified through these studies suggest that teams in crisis situations engage in less structured communication (David and Schraagen 2018; Stachowski et al. 2009) but return to systematic communication after a period of time elapses (Dunn et al. 2002), have fewer back-and-forth exchanges (Davis et al. 2016; Stachowski et al. 2009), and shift the focus of their communication over time (van der Haar et al. 2017). Our research contributes to this corpus of evidence through our holistic comparisons of teams communicating under different time constraints. Moreover, unlike any of the identified studies, we examine how the combination of time availability and media differences may influence communication patterns, and ultimately team performance.

What Is Included in this Book?

Overview of Chapters

Our aim in this book is to examine the longitudinal team communication exchanged among team members collaborating under different working conditions. Through this chapter we have developed a case for why this type of research is necessary, introduced the theoretical frameworks guiding our efforts, and described the different working conditions we consider.

In Chap. 2, we present the results of a structured literature review focused on longitudinal team communication. This review highlights the breadth of the existing research across the fields where the studies were published; the research designs employed; the way in which data were collected, coded, and manipulated; the analytical and data visualization techniques used; and findings that contribute to the literature on team science.

In Chap. 3, we provide an overview of the study conducted to collect the data examined in this book. In addition to descriptions of our sample and research design, we provide an extensive overview of how the data were coded and operationalized for the analyses that are covered in the subsequent chapters.

The next four chapters are organized around Bartunek and Woodman's (2015) dimensions of organizational change, as they facilitate a holistic examination of the change being accomplished by the teams as they complete their assigned tasks. Moreover, this investigation builds in complexity from simple visualization and statistical analyses to more complex and nuanced examinations of the relationships among topics.

In Chap. 4, we our focus is on the rhythm of communication, where rhythm depicts the repetition of cycles within the teams' conversations. Recurrence quantification analysis and linear mixed modeling are the analytical techniques used to assess communication rhythm. Comparisons are drawn across working conditions and between high and low performing teams. The research questions associated with rhythm are:

RQ4.1. Do teams have a rhythm in their communication over time?
RQ4.2. Does the rhythm of communication differ depending on the time they have available (i.e., unrestricted time versus time pressured)?
RQ4.3. Does the rhythm of communication differ depending on the communication media available to them (i.e., distributed versus in person)?
RQ4.4. Does the rhythm of communication differ depending on the cost performance achieved?

In Chap. 5, we examine communication timing and pacing. Timing is operationalized as the frequency with which teams discuss the topics of interest and pacing refers to the complexity of the transitions among topics. Network diagrams visually representing the frequencies with which topics are addressed. The transitions among topics, in conjunction with simple statistics, are used to compare communication timing and pacing across quartiles and among teams working in different contexts. The research questions associated with timing are:

RQ5.1. Does the frequency with which teams discuss different topics change over time?
RQ5.2. Does the frequency with which teams discuss different topics vary depending on the time they have available to complete their task (i.e., unrestricted time versus time pressured)?

RQ5.3. Does the frequency with which teams discuss different topics vary depending on the communication media available to them (i.e., distributed versus in person)?

The research questions associated with pacing are:

RQ5.4. Does the complexity of team conversations change over time?
RQ5.5. Does the complexity of team conversations vary depending on the time they have available to complete their task?
RQ5.6. Does the complexity of team conversations vary depending on the communication media available to them?

In Chap. 6, we consider polyphony to understand how team conversations align and overlap over time. Analytical techniques from event history analysis are used; specifically, we employ proportional hazard models and Andersen-Gill intensity models. Communication media and team performance are included in the models; unrestricted and time pressured teams are investigated independently. The research questions associated with polyphonic patterns among initial topic shift points (i.e., the points at which meaningful conversations about specific topics are completed or action is initiated) over time are:

RQ6.1. Do topic shift points interrelate to create polyphonic patterns over time?
RQ6.2. Do polyphonic patterns differ depending on the time teams have available (i.e., unrestricted time versus time pressured)?
RQ6.3. Do polyphonic patterns differ depending on the communication media available to teams (i.e., distributed versus in person)?
RQ6.4. Do polyphonic patterns relate to team performance?

The research questions investigating repetition in polyphonic patterns are:

RQ6.5. Do teams repetitively iterate through meaningful conversations about mission analysis and tactical strategy over time?
RQ6.6. What triggers these repetitive iterations to create polyphonic patterns over time?
RQ6.7. Do repetitive iterations within the polyphonic patterns differ depending on the time teams have available?
RQ6.8. Do repetitive iterations within the polyphonic patterns differ depending on the communication media available to teams?
RQ6.9. Do repetitive iterations within the polyphonic patterns relate to team performance?

In Chap. 7, we shift our attention to the sequences describing how topics are discussed. Hierarchical cluster analysis and dendrograms are used to assess communication sequences. Teams are partitioned by context conditions and performance for comparisons. The research questions associated with sequences are:

RQ7.1. Do teams have sequential patterns in their communication over time?
RQ7.2. Does the sequence of communication differ depending on the time teams have available (i.e., unrestricted time versus time pressured)?

RQ7.3. Does the sequence of communication differ depending on the communication media available to the team (i.e., distributed versus in person)?
RQ7.4. Do sequential patterns in team communication relate to cost performance?

Finally, Chap. 8 provides us with an opportunity to draw comprehensive conclusions about team communication across our results from the various analytical and visualization approaches employed. We also take the opportunity to suggest strategies for practitioners to enhance team communication and identify next steps for researchers.

Methodological Appendices

At the ends of Chapters 3 through 7, methodological appendices are provided. The purpose of these appendices is to help researchers employ the techniques in their own projects. The step-by-step instructions are written in accessible language that can easily be followed by other scholars. The appendix contents by chapter are:

- Introduction to the example dataset used throughout the appendices (Chap. 3)
- Recurrence quantification analysis and linear mixed modeling (Chap. 4)
- Network diagrams and social network density measures (Chap. 5)
- Proportional hazards models and Andersen-Gill intensity models from event history analysis (Chap. 6)
- Hierarchical cluster analysis and dendrograms (Chap. 7)

Managerial Minutes

In each chapter, we translate our research results into strategies that practitioners may find useful. Our intent is to formulate these strategies for immediate implementation by providing suggestions about, for example, ordering activities to be productive on the task (e.g., agenda setting), setting expectations about the duration of activities (e.g., meeting management), capitalizing on the relationships among topics (e.g., planning what needs to be done before jumping in to do it) and certain topics or patterns that signal a need for interventions (e.g., helping teams reach consensus about how they will approach their task).

Why Should You Read This Book?

We have prepared this book with both researchers and practitioners in mind. The results provide a thorough picture of longitudinal team communication because of the use of multiple analysis techniques. Through our approach we are able to make apples-to-apples comparisons of teams working under different constraints.

This holistic approach provides a view of team communication that would be infeasible with data from different sources. Isolating differences across communication media, the time available to complete the task, and the team's experience working together enhances our understanding of how these conditions may impact team dynamics. We then translate these results into practical takeaways.

Researchers

Our primary contribution is to the limited, existing research on longitudinal team communication. We employ a multidisciplinary, team process perspective to present a localized, objective event-based investigation of team communication. Moreover, we employ multiple methodologies to triangulate and reconcile insights into how communication progresses over time, may be influenced by time availability and working conditions, and may impact performance.

Our research is firmly grounded in existing theory from at least three perspectives. First, the research is positioned within the existing, limited research on team communication. Second, we draw guidance from the greater literature on team dynamics. More specifically, we build upon the growing body of evidence testing Marks and colleagues' (2001) recurring phase model of team processes by providing empirical evidence describing the temporality, interdependence, and performance implications of the transition and action processes they proffer. Third, our comprehensive examination of team communication is structured around Bartunek and Woodman's (2015) framework for exploring organizational change, which we borrow from the organizational development literature. Bartunek and Woodman suggest that change occurs across five temporal dimensions (rhythm, timing, pacing, rhythm, polyphony, and sequence) and advocate using their framework beyond organization-level change. We apply their framework at the team level since communication provides a lens through which team processes can be examined and these team processes govern the actions required to change from the current state to a specified future state (e.g., convert a concept for a new product into the production of a new product). Thus, our methodological approaches and findings may be germane for both the team dynamics and organizational development communities. These communities span a myriad of disciplines including, but not limited to cognitive science, communication, human factors, industrial engineering, organizational behavior, operations management, psychology, sociology, and strategy.

Our book may also become a useful methodological resource for researchers at all levels of proficiency. Not only do we draw theoretically-driven insights across the results obtained from multiple methodological approaches, we also include step-by-step guidance to help scholars employ these techniques in their own projects. These methodologies, while applied to team communication for our purposes, are applicable to a host of dynamic human behaviors both within and outside the team domain. Researchers outside the team domain may find these approaches applicable for studying topics such as individual-level decision making, geopolitical negotiation, health-promoting behaviors, or the history of labor movements.

Practitioners

Practitioners who manage or serve on teams may find our research-based recommendations useful. Throughout our book, we offer insights about how team communication occurs under varying circumstances. By learning about how these insights were derived through explanations of our research approach, practitioners may develop a deeper understanding of team dynamics. More importantly, we translate these insights into concrete suggestions that can be implemented immediately. These suggestions target issues ranging from setting meeting agendas to identifying when teams need to shift from iterating about multiple topics to focusing on specific issues to determining when certain topics should be discussed and in what order.

Managerial Minutes

Communication is often haphazard as teams collaborate to accomplish their missions. Throughout this book, we offer recommendations for more purposeful communication through these Managerial Minutes. Our intention is to identify concrete steps that team leaders and members can immediately enact to create small changes with potentially greater impact.

Team communication will differ due to teamwork conditions.
Consider the circumstances influencing how your team communicates. We examine the communication media teams use and the time available to them, but many more variables may influence collaboration. These variables include formal and informal hierarchies among team members, the team's size, familiarity among members, among so many others. Every team is different, but simply thinking about the uniqueness may help mitigate communication pitfalls.

Team communication needs to be managed.
A good start along this path is to be conscious of the communication going on around you and manage it. For example, if someone makes a good point that is important, but tangential to the current conversation, add it to a "parking lot" and return to it later to ensure the current discussion is not abandoned.

Chapter 2
Structured Literature Review

As highlighted in Chap. 1, limited team research has focused on how communication evolves over time. To showcase what has been accomplished, a structured literature review was conducted and is described in this chapter. The purpose of the review is to (1) highlight the variety of approaches available for researchers to gather, code, analyze, and visualize longitudinal team communication data and (2) demonstrate the breadth of knowledge about team interactions that can be obtained by implementing such approaches.

Literature Search Procedures

Team research is published across a wide array of fields. Therefore, the search strategy was designed to target as many of those fields as possible. To that end, the following databases were searched (total number of articles identified after duplicates were removed are in parentheses):

- ABI Inform Collection (303 unique articles)
- EBSCO Host Databases (565 unique articles)
  - Academic Search Premier
  - Applied Science and Technology Full Text
  - CINAHL
  - Communication and Mass Media Complete
  - Ergonomics Abstracts
  - MEDLINE with full text
  - Psych Articles
  - Psych Info
  - Social Sciences Full Text

S. McComb, D. Kennedy, *Computational Methods to Examine Team Communication*, Computational Social Sciences,
https://doi.org/10.1007/978-3-030-36159-4_2

- IEEE Xplore (62 unique articles)
- JSTOR (8 unique articles)
- Web of Science: Core Collection (108 unique articles)

The search string was TX = ("team proces*" OR "team communica*") AND TX = (longitudinal OR pattern) AND AB = communication. This string was used in all databases; minor variations were necessary based on the database search structure. These variations were done in a way to err on the side of caution by, for example widening the search in Web of Science; specifically, this database does not have an option to search the abstract so all fields were searched instead. Only scholarly publications written in English and published in peer reviewed journals were searched for the years 1999–2019. This search strategy resulted in 1046 articles identified, and after duplicates were removed 931 articles were available for a title and abstract review.

Guiding the review were the inclusion and exclusion criteria. As deduced from the search criteria, only peer-reviewed scholarly journal articles written in English between 1999 and 2019 were included. Furthermore, inclusion required that the articles incorporate team communication *content* and/or *action* over time and offer data-driven results that inform team science. Excluded articles focused on use of communication media over time, development and testing of methods to analyze communication longitudinally, or investigation of frequency counts over time without longitudinal statistical analysis or visualization; intervention studies were also excluded.

After the title and abstract review, 91 articles remained for full article review. The same inclusion and exclusion criteria were applied. Through this screening, 27 articles were retained for inclusion in this chapter. Through the review of these articles, we became aware of significant contributions to this area of research by two authors, Kauffeld and Lehmann-Willenbrock, that were not identified through our search parameters. These omissions caused us to identify other authors with multiple articles included in our review (e.g., Cooke, Gorman, Miller). We conducted a search of these authors' articles and identified an additional 13 for inclusion. Upon reviewing these articles to ascertain why they may not have been identified in the original search, we determined that communication was described by various terms including behaviors, coordination, interactions, and/or utterances; longitudinal effects were identified by terms other than patterns such as sequences; and/or teams were identified as groups. Finally, six articles were identified through a more traditional legacy search, such as adding articles of which we were aware prior to conducting the review or identifying articles through our review. Although our search strategy did not identify these articles, we believe that the final list of included articles is comprehensive and, more importantly, achieves our purpose in highlighting the variety of approaches available to examine team communication over time.

The resulting 46 articles are summarized in Table 2.1, where we highlight information about the research aim; sample population and task; communication data, coding, and theoretical framework; analysis technique(s) and how data were used; visualization technique(s); and findings. Only information relevant to our interest in

Table 2.1 Summary of articles included in structured literature review

Authors, journals, and publication date	Research aim	Sample, population, and task	Communication data, coding, and theoretical framework	Analysis technique(s) and how data were used	Visualization technique(s)	Findings
Aufegger, Bicknell, Soane, Ashrafian, & Darzi *BMC Medical Research Methodology* 2019	Examine feasibility of applying signal processing and machine learning to understand teamwork in healthcare	28 3-5 member teams of clinical and non-clinical healthcare professionals in the United Kingdom participated in role-play exercise	*Data:* Interactions among team members from video recording and communication transcripts from audio recording *Coding:* Manual coding of communication transcripts to identify team processes *Theoretical Framework:* Bales' (1950) interaction process analysis model	*Technique:* Recurrence quantification analysis; recurrence rate, determinism, and entropy are reported *Data Use:* Coded communication strings were statistically analyzed	None	As recurrence rates (i.e., conversation stability) increased and determinism (i.e., predictability) decreased, perceived social support increased
Barth, Schraagen & Schmettow *Ergonomics* 2015	Examine communication patterns within a paediatric cardiac surgical team to diagnose performance	40 live procedures completed by 1 8-9 member paediatric cardiac surgical team in the Netherlands; team conducted paediatric cardiac surgeries	*Data:* Communication among team members documented in real time by two trained observers *Coding:* Manual coding about: (1)who talks to, works with, and reports to whom (during observation); (2)types of communication utterances such as answer, command, request (on communication transcripts) *Theoretical Framework:* Santos et al.'s (2012) categorisations	*Technique:* Social network analysis; network diagrams were constructed for the seven phases of the surgical procedure; density, reciprocity, degree centralisation, betweenness centralisation, closeness centralisation by phase are reported *Data Use:* Interaction among team members was aggregated in the seven phases of the surgical procedure; communication content was used to interpret differences across types of phases (surgical phase or transition between surgical phases)	Social networks across different phases and for low and high complexity surgeries	Information sharing (network density) and closed-loop communication (reciprocity) were lower during surgical phases than when the team was transitioning from one phase of the surgery to the next; these findings were more pronounced in complex surgeries
Bourbousson, R'Kiouak & Eccles *European Journal of Work and Organizational Psychology* 2015	Examine shared awareness among team members embedded in a dynamic network	2 male basketball teams in France; one expert and one novice team played basketball	*Data:* Team members' verbal play-by-play descriptions of their actions, perceptions, feelings, and thoughts were captured as they viewed the first eight minutes of the novice game and ten minutes of the expert game *Coding:* Manual coding of each team member's behaviors combined with communicated play-by-play descriptions *Theoretical Framework:* Theureau's (2003) course-of-action analysis	*Technique:* Social network analysis; network diagrams were constructed for each discrete meaningful unit; indegree centrality, outdegree centrality, betweenness centrality, reciprocity, transitivity, density, and centralization for each discrete meaningful unit at the micro-, meso-, and macro-levels are reported *Data Use:* Content codes were used to break activity into discrete meaningful units for longitudinal analysis	Plots over time of (1) social network analysis indicators (e.g., reciprocity, density) and (2) running correlations between density and transitivity	Dyads transitioned into triads via emergent shared awareness. For expert teams, static indicators suggest individualistic play, whereas longitudinal analyses suggest collaborative behavior over time. This pattern of collaborative behavior was more stable in expert teams than novice teams over time

(continued)

Table 2.1 (continued)

Chiocchio *Project Management Journal* 2007	Identify differences in how high and low performing teams discuss their tasks and how to coordinate.	34 2-6 person teams of first-year psychology students in Canada created and validated an assessment instrument based on psychometric theory	*Data:* Team discussion board postings *Coding:* Manual coding to identify postings reflecting issues or solutions associated with the task or coordination *Theoretical Framework:* None	*Technique:* Time series regression analysis using auto regressive integrated moving average (ARIMA) *Data Use:* Coded communication strings were used to create plots and statistically analyzed	Plots over time, where task issues and solutions or coordination issues and solutions were plotted for high or low performing teams	High performing teams, compared to low performing teams (1) engaged in significantly more task and coordination communication leading up to milestone deadlines and (2) increased significantly the number of exchanges about task and coordination solutions two weeks before the deadline signifying a last surge to self-organize around task-related activity
Chiu & Lehmann-Willenbrock *Group Dynamics* 2016	Investigate how social metacognitive actions influence idea generation and justification	20 4-person teams of 9th grade students in the United States solved an algebra problem	*Data:* Transcripts of video recordings *Coding:* Manual coding of 7 actions (e.g., justification, agree, disagree rudely) *Theoretical Framework:* None	*Technique:* Statistical discourse analysis *Data Use:* Coded communication strings were statistically analyzed	None	Rude disagreements led to (1) less creativity when it triggered more rude disagreement or (2) more creativity when it followed a wrong idea or in groups able to solve the problem; exchanges made by students with higher mathematics grades were often followed by justifications
David & Schraagen *Cognition, Technology & Work* 2018	Identify changes in communication dynamics when a team experiences an unexpected situation	3-person airplane cockpit crew employed by a French airline flew an airplane	*Data:* Official flight transcript communication among captain and two co-pilots *Coding:* Manual coding of dyadic communication noting time, communicator, receiver, and whether the crew was in a normal or emergency phase *Theoretical Framework:* None	*Technique:* Relational event modeling; normal and emergency phases data examined independently; fixed effects, preferential attachment, triadic effects, persistence, recency, and participation shifts (reciprocation, hand off, persistence, and attraction) were included in models *Data Use:* Coded communication strings were statistically analyzed	Plot of parameter estimates and approximate 95% confidence intervals for normal and emergency phase results	Pattern changes were identified between the normal phase and the emergency phase, with standardized communication procedures prevalent in the normal phase and more localized communication patterns that rely on immediately preceding events in the emergency phase
Davis, Jones, Crowell-Huhnberg, O'Keeffe, Boyle, Klainer, Smink, & Yule *Surgery* 2016	Examine use of closed-loop communication in an emergency situation	7 interprofessional teams comprised of 2 nurses, 2 attending anesthesiologists, and 2 attending surgeons in the United States participated in a simulation training exercise for post-anesthesia care	*Data:* Audio and video recordings *Coding:* Manual coding to identify messages that were (1) commands, questions, observations, or suggestions, (2) directed (including from and to) or non-directed, and (3) check back or closed-loop communication *Theoretical Framework:* None	*Technique:* Zero inflated Poisson and mixed-effects negative binomial regression *Data Use:* Coded communication strings were used to create plots and statistically analyzed	Overlying kernel density plots	Directed communication increased the likelihood of receiving a check back communication; but the effect was only for periods preceding an emergency situation since during an emergency situation they disappeared

Authors, journals, and publication date	Research aim	Sample, population, and task	Communication data, coding, and theoretical framework	Analysis technique(s) and how data were used	Visualization technique(s)	Findings
Dunn, Lewandowsky & Kirsner *Applied Cognitive Psychology* 2002	Examine dynamics of verbal exchanges among key members of an emergency management team during an emergency	3 2-officer teams from a fire and rescue service in Australia participated in a simulation training exercise focused on response to an emergency chemical spill	*Data:* Transcripts of audio recordings from three communication channels (one officer-to-officer and two officer-to-field) *Coding:* Manual coding of when specific events occurred that were observed in at least two of the three teams and required a significant response from an officer *Theoretical Framework:* None	*Technique:* Time series regression with an autoregression term of lag one included in each analysis *Data Use:* Coded communication strings were used to calculate weighted sums of events during sequential five-minute windows	Plot of critical events and best fitting regression line; plots of observed event frequencies, moving averages, and predictions for five-minute intervals of each communication channel	Timing of verbal exchanges was correlated with timing of critical events and followed a 30-minute temporal cycle
Gerpott, Lehmann-Willenbrock, Voelpel, VanVugt *Academy of Management Journal* 2019	Investigate the relationship between emergent leadership and task-, change-, and relations-oriented behaviors	42 3-5 person self-managed project teams of international university students in Germany developed and proposed a solution to a problem in an automobile manufacturing company	*Data:* Video recordings *Coding:* Manual coding of task-, change-, and relations-oriented behaviors *Theoretical Framework:* None	*Technique:* Multilevel modeling *Data Use:* Aggregated counts of the three types of coded behaviors by time period were statistically analyzed	None	Change-oriented communication became less relevant over time, relations-oriented communication became increasingly important, and task-oriented communication was stable over time
Gorman, Amazeen & Cooke *Nonlinear Dynamics, Psychology, and Life Sciences* 2010	Capture momentary fluctuations in team coordination	45 3-person teams in the United States navigated and flew reconnaissance photography missions (11-12 targets per mission) on an unmanned aerial vehicle (UAV) simulator; two sessions with either a short or long interval between sessions; returning teams had either the same or mixed membership for the second session	*Data:* Real-time observations of team member communication behaviors *Coding:* Observer ratings of coordination components (operationalized as navigator-provided information, pilot and photographer negotiation, and photographer feedback) *Theoretical Framework:* None	*Technique:* Dynamical analyses including attractor reconstruction, Hurst analysis, and stability analysis *Data Use:* Coordination components were used to calculate a coordination order parameter for each target	Plots of coordination components over time with indications of when perturbations were introduced	Intact and mixed teams had different coordination dynamics, with mixed teams being more adaptive and stable than intact teams

(continued)

Table 2.1 (continued)

Gorman & Cooke *Journal of Experimental Psychology: Applied* 2011	Identify changes in the retention of team cognition and processes when team membership changes	40 3-person teams in the United States navigated and flew reconnaissance photography missions (11-12 targets per mission) on a UAV simulator; two sessions were run with either a short or long interval between sessions and returning teams had either the same or mixed membership for the second session	*Data:* Real-time observations of team member behaviors; automated calculation of communication communication frequency and duration per mission *Coding:* Observer ratings of team processes (operationalized as an assessment of communication accuracy, timeliness, and quality) per target *Theoretical Framework:* None	*Technique:* Repeated measures ANOVA for team process; multivariate MANCOVA for communication frequency and duration *Data Use:* Aggregated counts of team process, communication frequency, and communication duration by mission were statistically analyzed	Plots of communication frequency and duration over time	Team process (i.e., communication accuracy, timeliness, and quality) improved and plateaued during Session 1; mixed teams' team processes improved over Session 2, but intact teams remained the same; in the second session, communication frequency significantly decreased early and duration decreased later
Gorman, Cooke, Amazeen & Fouse *Human Factors* 2012	Examine communication patterns when team membership changes	37 3-person teams in the United States navigated and flew reconnaissance photography missions (11-12 targets per mission) on a UAV simulator; two sessions with either a short or long interval between sessions; returning teams had either the same or mixed membership for the second session	*Data:* Automatically generated record of who was talking, when, and for how long *Coding:* None *Theoretical Framework:* None	*Technique:* Recurrence quantification analysis; determinism and pattern information are reported *Data Use:* Communication strings were statistically analyzed	Recurrence plots; plots of determinism and pattern information across missions	Mixed teams engage in less rigid communication and exhibit more unique patterns in their exchanges than intact teams

Authors, journals, and publication date	Research aim	Sample, population, and task	Communication data, coding, and theoretical framework	Analysis technique(s) and how data were used	Visualization technique(s)	Findings
Gorman, Hessler, Amazeen, Cooke & Shope *Ergonomics* 2012	Conduct real-time analyses of dynamic team communication	1 3-person team in the United States navigated and flew 5 reconnaissance photography missions (9 targets per mission in missions 1-4 and 20 targets in mission 5) on a UAV simulation	*Data:* A three-column multivariate time series (one column for each members' communication activity) representing talking/not talking at 1 second intervals *Coding:* None *Theoretical Framework:* None	*Technique:* Confidence intervals and regression *Data Use:* Communication strings were statistically analyzed	Plots over time of Lyapunov exponents and least-squares slopes of variability	The team established an intrinsic, dynamic flow of information among team members that was recovered after the team dynamics had been perturbed by an external request that had to be fulfilled
Gorman, Martin, Dunbar, Stevens, Galloway, Amazeen & Likens *Human Factors* 2016	Investigate cross-level effects between brainwaves and team communication	7 6-person submarine officer teams (4 experienced and 3 inexperienced teams) participated in submarine piloting and navigation simulation training exercise that included briefing, completing a training scenario, and debriefing	*Data:* Latent semantic analysis (LSA) input matrix of frequency co-occurrence of unique words by unique paragraphs, constructed from archival materials describing submarine communication and the team training exercise transcripts *Coding:* Each utterance was represented as a vector length calculated as the amount of speech weighted by the content of the discourse *Theoretical Framework:* None	*Technique:* ANOVA, hierarchical cluster analysis, cross-correlation analysis *Data Use:* Vector length was aggregated for each team at each training segment; vector lengths were used to calculate LSA cosine between all possible pairs of transcripts as a function of training segment and team experience	Plots of mean vector lengths by training segment, cross-correlation functions over time for one more and one less experienced team, peak cross-correlations by team, absolute peak cross-correlation for each training segment, peak cross-correlation lag for each training segment and experience level	Teams approach communication similarly for briefings and scenario completion by being terse and domain specific, but shift to more lengthy discourse during debriefing; more experienced teams exhibited changes in communication that preceded changes in neural patterns
Grand, Braun, Kuljanin, Kozlowski & Chao *Journal of Applied Psychology* 2016	Evaluate team knowledge emergence from a process-oriented theory perspective	None	*Data:* Communication data is generated within a computational model based to exchange information among team members as they generate shared knowledge *Coding:* None *Theoretical Framework:* None	*Technique:* Computational modeling and simulation *Data Use:* Communication data is simulated under varying conditions by manipulaing team member information processing skills, communication skills and degrees of specialization	None	Knowledge emergence is enhanced by (1) improving the rate at which team members process information and (2) promoting more frequent and equal distribution of communication across team members

(continued)

Table 2.1 (continued)

Gundrosen, Andenæs, Aadahl & Thomassen *Scandinavian Journal of Trauma, Resuscitation and Emergency Medicine* 2016	Investigate interrelationships between communications and actions during medically critical phases of teamwork	5 acute medical response teams, with an average of 6 persons per team, in Norway participated in a simulation training exercise focused on acute medical response team routines for treating critically ill patients	*Data:* Transcripts of video recordings augmented with notes from field observations that include verbal exchanges, interaction activities, and medical activities *Coding:* Manual coding that (1) identified key activities phases and associated sub-phases, (2) coded specific medical and interactional utterances and actions to construct a sequential organization of communication, and (3) pinpoint teamwork phases that are medically and/or communicatively critical *Theoretical Framework:* None	*Technique:* Qualitative discourse analysis *Data Use:* Coded communication and action strings were thematically analyzed	None	Three key activity phases were identified; interpreting and sharing information with everyone involved played a critical role in ensuring patient safety, particularly in the second phase when team membership was expanding based on the care needs of the patient
Havakhor & Sabherwal *Journal of Management Information Systems* 2018	Understand how information technology artifacts affect transactive processes and team performance	62 6-person virtual knowledge teams recruited from Amazon's Mechanical Turk designed web pages	*Data:* Communication logs from online collaboration interface *Coding:* Manual coding of 7 categories of activities (e.g., information events, action events) *Theoretical Framework:* Marks et al. (2001) recurring phase model	*Technique:* Optimal matching algorithm; *t*-tests *Data Use:* Coded communication strings were statistically analyzed to calculate the intersequence distances between teams	None	Virtual knowledge teams exhibit communication patterns similar to those of teams with comparable (1) performance levels, (2) network densities, and (3) reputation signals
Hoogeboom & Wilderom *Group & Organization Management* 2019	Identify effective patterns of team interaction	96 teams, with an average of 13 persons per team, from one large public-sector organization in the Netherlands conducted a team staff meeting	*Data:* Video recordings *Coding:* Manual coding of 18 mutually exclusive micro-behaviors (e.g., task monitoring, interrupting), which were transformed into 4 behavioral meta-categories that align with leadership theory and 6 micro-behaviors that did not align with the meta-categories (e.g., humor, agreeing, providing positive feedback) *Theoretical Framework:* Hoogeboom & Wilderom's (2015) behavioral categories	*Technique:* Hierarchical multiple regression analysis *Data Use:* The pattern recognition algorithms embedded in Theme software were used to identify behavioral patterns over time in the coded communication strings; aggregated values representing the total number of times patterns occurred, the numbers of unique patterns, and speaker switches were calculated and used in statistical analysis	None	Frequently recurring interaction patterns occurring during non-routine tasks may reduce the perceived amount of information shared, which in turn may lower team effectiveness; the numbers of unique patterns had no significant relationship with information sharing or effectiveness; speaker switching may be beneficial for perceived information sharing and team effectiveness for all task types

Authors, journals, and publication date	Research aim	Sample, population, and task	Communication data, coding, and theoretical framework	Analysis technique(s) and how data were used	Visualization technique(s)	Findings
Kauffeld & Meyers *European Journal of Work and Organizational Psychology* 2009	Investigate cycles of complaining and solution-oriented behaviors	33 5-7 person blue collar work groups from three medium sized companies in Germany identified possible improvement opportunity solutions and planned for implementation	*Data:* Transcripts of video recordings *Coding:* Manual coding of 44 competencies *Theoretical Framework:* Kauffeld's (2006) Cassel competence grid	*Technique:* Lag sequential analysis *Data Use:* Coded communication strings were statistically analyzed	None	Complaining begets more complaining, supporting statements may reinforce the sequence, and structuring statements may break the cycle; solution-oriented discussions do not require supporting statements, but when supporting statements do occur they encourage additional conversation about problem identification or description statement but not examinations of disadvantages with or alternative solutions
Kennedy & McComb *Journal of Applied Psychology* 2014	Determine when team process shifts may enhance performance	63 3-person teams of undergraduate students in the United States completed a scheduling assignment problem	*Data:* Transcripts of audio recordings *Coding:* Manual coding of 5 processes (e.g., tactical strategy, action) *Theoretical Framework:* Marks et al.'s (2001) recurring phase model	*Technique:* Computational modeling, genetic algorithm optimization, and simulation *Data Use:* Coded communication strings were used to (1) train neural networks and (2) construct and validate a computational model for virtual experimentation	None	Teams may benefit from (1) discussing what they have been tasked with doing, how they are going to do it, and who is going to do what before the midpoint of collaboration, (2) waiting to initiate any action to complete the task until after the midpoint of collaboration, and (3) determining how they are going to approach their task before completing any discussions of what they have been tasked with doing
Klonek, Quera, Burba & Kauffeld *Group Dynamics* 2016	Identify differences in communication patterns between meetings held during the first and second halves of the project life cycle	24 meetings held by 1 team, with an average of 10 members per meeting, from the automotive industry conducted a team meeting	*Data:* Video recordings *Coding:* Manual coding of message types (clarifying, procedural suggestion, summarizing) *Theoretical Framework:* Kauffeld & Lehmann-Willenbrock's (2012) act4teams coding scheme	*Technique:* Lag sequential analysis *Data Use:* Coded communication strings were statistically analyzed	None	Positive procedural communication may reduce dysfunctional communication patterns; Negative action-oriented communication was only observed during the second half of the team's project life cycle
Kolbe, Grote, Waller, Wacker, Grande, Burtscher & Spahn *Journal of Applied Psychology* 2014	Identify differences in how high and low performing teams engage in team member monitoring and talking to the room	27 3- to 4-person ad hoc anesthesia teams, comprised of registered nurses and both resident and attending physicians, in Switzerland performed inductions of general anesthesia	*Data:* Video recordings *Coding:* Manual coding of task-relevant "talk to the room" messages (i.e., communication addressed to the group versus a specific team member) was coded as action-related or information-related *Theoretical Framework:* None	*Technique:* Lag sequential analysis *Data Use:* Coded communication strings were statistically analyzed	None	High performing teams, compared to low performing teams, (1) engaged in more autocontingencies, where both action- and information-related talk to the room messages were more likely to be followed by messages of the same type and (2) were less likely to provide instructions after delivering action-related talk to the room messages

(continued)

Table 2.1 (continued)

Lee & Jamison *Journal of Educational and Psychological Consultation* 2003	Examine how functional behavioral assessments are incorporated into the student assistance team process	1 7-member student assistance team in the United States reviewed cases of students referred to the student assistant team	*Data:* Transcripts of video recordings *Coding:* Manual coding of team processes *Theoretical Framework:* Bergan & Tombari's (1975) consultation analysis record (CAR)	*Technique:* Qualitative visual analysis *Data Use:* Percentages of each CAR process were calculated for each of the three team meetings	Graphs representing the percentage of messages by CAR code were plotted for each meeting	Early meetings focused on descriptive information exchange and then shifted to feedback on data collected, hypothesis generation, and intervention planning; observation-related statements peaked during the second meeting and the proportions of planning statements increased across all three meetings
Lehmann-Willenbrock & Allen *Journal of Applied Psychology* 2014	Determine how humor patterns develop and influence team outcomes	54 teams, with an average of 6.5 persons per team, from two manufacturing organizations in Germany conducted a team meeting	*Data:* Video recordings *Coding:* Manual coding of message types (clarifying, procedural suggestion, summarizing) with additional emphasis on humor statements *Theoretical Framework:* Kauffeld & Lehmann-Willenbrock's (2012) act4teams coding scheme	*Technique:* Lag sequential analysis and linear regression analysis *Data Use:* Coded communication strings were statistically analyzed to identify patterns; aggregated frequency counts of patterns were used in statistical analyses	None	Humor patterns (1) trigger functional team behaviors and (2) positively relate to performance immediately and over time; the positive relationship between humor patterns and performance exists when team members are more secure in their jobs
Lehmann-Willenbrock & Allen *International Journal of Business Communication* 2017	Examine team communication patterns when members are late versus everyone is on time	32 4-6 person teams of undergraduate students in the United States participated in a laboratory experiment, where one set of teams started on time and another waited for 10 minutes before being told that the team member they were waiting for was not coming	*Data:* Video recordings *Coding:* Manual coding of message types (clarifying, procedural suggestion, summarizing) *Theoretical Framework:* Kauffeld & Lehmann-Willenbrock's (2012) act4teams coding scheme	*Technique:* Lag sequential analysis *Data Use:* Coded communication strings were statistically analyzed	None	Communication patterns were significantly different for teams starting late versus those starting on time; in teams starting late (1) solution and problem analysis statements did not receive support, (2) members did not argue for solutions, (3) problems led to complaining, and (4) root cause analysis was less likely to occur
Lehmann-Willenbrock, Allen & Kauffeld *Journal of Applied Communication Research* 2013	Investigate using strategically structured communication in improve team meetings	59 5- to 7-person teams from a broad array of industries in Germany conducted a team meeting	*Data:* Video recordings *Coding:* Manual coding of message types (clarifying, procedural suggestion, summarizing) *Theoretical Framework:* Kauffeld & Lehmann-Willenbrock's (2012) act4teams coding scheme	*Technique:* Lag sequential analysis *Data Use:* Coded communication strings were statistically analyzed	Plots of common sequences	Procedural meeting behaviors (1) precede supporting statements, (2) promote proactive communication, (3) inhibit dysfunctional behavior, and (4) when distributed among team members, relate to meeting satisfaction

Authors, journals, and publication date	Research aim	Sample, population, and task	Communication data, coding, and theoretical framework	Analysis technique(s) and how data were used	Visualization technique(s)	Findings
Lehmann-Willenbrock, Allen & Meinecke *Group Processes & Intergroup Relations* 2014	Compare meeting cultures in Germany and the United States	15 3-4 person teams of predominately undergraduate students in Germany and 15 5-person teams of undergraduate students in the United States completed a task requiring a consensus decision	*Data:* Video recordings *Coding:* Manual coding of message types (clarifying, procedural suggestion, summarizing) *Theoretical Framework:* Kauffeld & Lehmann-Willenbrock's (2012) act4teams coding scheme	*Technique:* Lag sequential analysis *Data Use:* Coded communication strings were statistically analyzed	None	Teams in Germany focused primarily on problems and procedures; teams in the United States focused on solutions
Lehmann-Willenbrock, Beck & Kauffeld *Communication Studies* 2016	Determine if emergent team roles can be identified through communication behaviors	59 5- to 7-person teams from a broad array of industry in Germany conducted a team meeting	*Data:* Transcripts of video recordings *Coding:* Manual coding of message types (clarifying, procedural suggestion, summarizing); idea longevity was assessed by determining how thoroughly, if at all, an idea was discussed at subsequent meetings *Theoretical Framework:* Kauffeld & Lehmann-Willenbrock's (2012) act4teams coding scheme	*Technique:* Hierarchical cluster analysis, correlation analysis, and *t*-test *Data Use:* Coded communication strings were used to (1) aggregate the frequency each team member exhibited specific behaviors and (2) qualitatively assess idea longevity	None	Teams with complainers had lower idea-longevity scores; teams with solution seekers had higher idea-longevity scores
Lehmann-Willenbrock & Chiu *Journal of Organizational Behavior* 2017	Examine how disagreements start and stop during team interactions	43 problem-solving teams, with an average of 6.2 participants per team meeting, from two medium sized firms in Germany conducted a team meeting	*Data:* Video recordings *Coding:* Manual coding of message types (clarifying, procedural suggestion, summarizing); idea longevity was assessed by determining how thoroughly, if at all, an idea was discussed at subsequent meetings *Theoretical Framework:* Kauffeld & Lehmann-Willenbrock's (2012) act4teams coding scheme	*Technique:* Statistical discourse analysis *Data Use:* Coded communication strings were statistically analyzed	None	Team-level conflict and productivity are not related to starting a disagreement or ending by agreeing, but individual-level problem-solving and off-tasks behaviors are; team viability was positively related to starting content disagreements

(continued)

Table 2.1 (continued)

Lehmann-Willenbrock, Chiu, Lei & Kauffeld *Group & Organization Management* 2017	Examine positivity within dynamic team interactions	43 problem-solving teams, with an average of 6.2 participants per team meeting, from two medium sized firms in Germany conducted a team meeting	*Data:* Video recordings *Coding:* Manual coding of message types (clarifying, procedural suggestion, summarizing); idea longevity was assess by determining how thoroughly, if at all, an idea was discussed at subsequent meetings *Theoretical Framework:* Kauffeld & Lehmann-Willenbrock's (2012) act4teams coding scheme	*Technique:* Statistical discourse analysis *Data Use:* Coded communication strings were statistically analyzed	None	Positivity may be inhibited by problem-focused utterances and promoted by (1) solution-focused utterances, which is enhanced by speaker switches, and (2) earlier positivity
Lehmann-Willenbrock, Meinecke, Rowold & Kauffeld *The Leadership Quarterly* 2015	Examine how and why transformational leadership may facilitate team functioning	30 continuous improvement process teams, with an average of 6.1 participants per team meeting, from two manufacturing organizations in Germany conducted a team meeting	*Data:* Video recordings *Coding:* Manual coding of message types (clarifying, procedural suggestion, summarizing) and codes were attributed to either the team leader or to members *Theoretical Framework:* Lehmann-Willenbrock et al.'s (2011) act4teams coding scheme	*Technique:* Lag sequential analysis *Data Use:* Coded communication strings were statistically analyzed	None	Solution-focused statements from the team leader (1) triggered solution-focused statements from the members and (2) inhibited members from making counterproductive statements
Lehmann-Willenbrock, Meyers, Kauffeld, Neininger & Henschel *Small Group Research* 2011	Examine how complaining and interest-in-change may be related to group mood	52 5-7 person blue-collar production work groups from two medium-sized firms in Germany identified possible improvement opportunity solutions and planned for implementation	*Data:* Transcripts of video recordings *Coding:* Manual coding of message types (clarifying, procedural suggestion, summarizing) *Theoretical Framework:* Kauffeld & Lehmann-Willenbrock's (2012) act4teams coding scheme	*Technique:* Lag sequential analysis *Data Use:* Coded communication strings were statistically analyzed	None	Group mood can be created and sustained through planning communication and cycles of complaining
Lingard, Reznick, Espin, Regehr & DeVito *Academic Medicine* 2002	Examine communication patterns among operating room team members	35 live surgical procedures (no information about how many times specific teams were observed) in the Canada conducted surgeries	*Data:* Ethnographic field notes recording communication-related activities *Coding:* Manual coding of communicative events, including the content, participants, and context *Theoretical Framework:* None	*Technique:* Qualitative inductive analysis *Data Use:* Coded communication events were thematically analyzed	None	Communication patterns indicate that surgical teams enact communication strategies to simultaneously achieve goals, minimize tension, and maintain social cohesion; tense communication patterns can be transmitted to novices

Authors, journals, and publication date	Research aim	Sample, population, and task	Communication data, coding, and theoretical framework	Analysis technique(s) and how data were used	Visualization technique(s)	Findings
McComb, Kennedy, Perryman, Warner & Letsky *Human Factors* 2010	Capture and compare temporal cognitive processes of distributed and face-to-face teams by analyzing communication	32 3-person teams of community college students in the United States developed a rescue plan for a simulated military scenario	*Data:* Distributed teams: communication logs from electronic chat; face-to-face: transcripts of audio recordings *Coding:* Manual coding of 13 mutually exclusive topics inductively developed and iteratively refined *Theoretical Framework:* None	*Technique:* Qualitative discourse analysis; event history analysis *Data Use:* Coded communication strings were used to identify patterns and statistically analyzed	Graphics depicting communication patterns of distributed and face-to-face teams for comparison	Mental model convergence points (i.e., topic shifts) (1) occur over time, (2) are temporally interdependent, (3) influence performance, and (4) differ for distributed and face-to-face teams
Miller, Scheinkestel & Joseph *Human Factors* 2009	Examine how care coordination unfolds through team communication in an intensive care unit	5 patient care teams, with revolving physician and nurse members, in Australia cared for critically ill patients	*Data:* Ethnographic field notes recording the date, time, conversation context, conversation content, speaker roles, and supporting artifacts; audio recording were used to ensure field note accuracy *Coding:* Manual coding of coordination phase, dyadic communicators, context (e.g., formal handoffs, informal telephone requests), time horizon for next act (e.g., immediately, in 12 hours), and artifacts supporting coordination (e.g., cardiac monitors, charts) *Theoretical Framework:* Klein's (2001) team coordination phases	*Technique:* Markov analysis *Data Use:* Coded communication strings were statistically analyzed	Markov diagrams depicting transitions within and between phases of care for formal and informal communication	Care coordination does not progress linearly through traditional phases
Miller, Weinger, Buerhaus & Dietrich *Human Factors* 2010	Examine the complexity of communication and care coordination in intensive care units	32 charge, fellow, resident, advanced practice, or registered nurses and the patient care team members with whom they interacted from one teaching medical center in the United States cared for critically ill patients	*Data:* Ethnographic field notes recording the time, speaker roles, and conversation topic and outcomes *Coding:* Manual coding of unit, roles, decision-making level (e.g., care coordination, patient care planning), coordination phase, and information space (e.g., diagnosis, goals, data) *Theoretical Framework:* Klein's (2001) team coordination phases; Miller's (2004) patient information spaces	*Technique:* Multiway contingency table analysis *Data Use:* Coded communication strings were statistically analyzed	Plots of patient information and team member roles across team coordination phases for the levels of decision making	Patient information spaces and role holders differ across team coordination phases and decision-making levels

(continued)

Table 2.1 (continued)

Olekalns, Brett & Weingart *The International Journal of Conflict Management* 2003	Examine patterns of negotiation strategy use in teams over time	36 4-person teams of MBA students in the United States and/or Australia participated in a negotiation role playing exercise	*Data:* Transcripts of video recordings *Coding:* Manual coding of 32 behavioral categories *Theoretical Framework:* Pruitt & Carnevale's (1982) social conflict categorization	*Technique:* WinPhaser; gamma statistical analysis *Data Use:* Coded communication strings were statistically analyzed	Example graphic of one team's phase map and phase diagram	Most teams (1) began their negotiations with common phase sequences (i.e., distributive phases preceded integrative phases and action preceded information) and (2) and changed only the orientation or the function of their approach at a time (versus both at the same time)
Parush, Kramer, Foster-Hunt, McMullan & Momtahan *Cognition, Technology & Work* 2014	Investigate similarities and differences across healthcare work contexts to ascertain the feasibility of knowledge transfer	25 surgical teams and 80 handoff teams in Canada conducted surgeries or completed patient handoff to colleagues on another unit or the next shift	*Data:* Field notes from real-time observations; handoffs were audio recorded *Coding:* Manual coding of verbal behaviors (e.g., asking question, providing answer) and contextual factors (e.g., speakers, procedure phase) *Theoretical Framework:* None	*Technique:* None *Data Use:* Coded communication strings were used to aggregate the frequency counts of behaviors by phase	Plots of communication frequency in each phase of surgery	Communication frequency followed a similar pattern over time across the two types of surgeries; no phases were identified in handoffs
Rapanta, Maina, Lotz & Bacchelli *Educational Technology Research & Development* 2013	Capture how design decisions are made	2 teams of instructional designers in Europe designed and produced a new course	*Data:* Transcripts of video recordings *Coding:* Manual coding of messages as design or communication messages; manual coding within these categories *Theoretical Framework:* None	*Technique:* Frequencies of increasingly long sequences within the communication strings were calculated to identify patterns *Data Use:* Communication strings were statistically analyzed	None	The majority of communication patterns are circular (e.g., method-problem-method) or repetitive (e.g., detail-detail-detail) versus linear (e.g., propose-explain-comment)
Schecter, Pilny, Leung, Poole & Contractor *Journal of Organizational Behavior* 2017	Connect event sequences to emergence in teams	55 4-person teams of military officers from the United States, Bulgaria, Norway, Sweden, and the Netherlands developed and executed a plan for a simulated military scenario	*Data:* Communication logs from online collaboration interface *Coding:* Automated coding of sender, received, and time of each communication event *Theoretical Framework:* Pilny et al.'s (2016) relational event framework	*Technique:* Hierarchical relational event modeling *Data Use:* Communication strings were used to calculate individual-level statistics representing the frequency, timing, and pattern of interactions	None	Sequences of interactions among team members emerge from individual-level behavior and influence team outcomes; static examinations of the relationships tested were not significant

Authors, journals, and publication date	Research aim	Sample, population, and task	Communication data, coding, and theoretical framework	Analysis technique(s) and how data were used	Visualization technique(s)	Findings
Schneider, Klünder, Kortum, Handke, Straube & Kauffeld *The Journal of Systems & Software* 2018	Examine the relationship between team communication and affect at the individual and team levels	32 3-5 person teams of undergraduate students in Germany completed a course project	*Data:* Video recordings *Coding:* Manual coding of message types (clarifying, procedural suggestion, summarizing) *Theoretical Framework:* Kauffeld & Lehmann-Willenbrock's (2012) act4teams coding scheme	*Technique:* Lag sequential analysis *Data Use:* Coded communication strings were statistically analyzed	None	Proactive statements had a positive impact on team-level affect, but only when followed by supportive statements
Seeber, Maier, & Weber *Group Decision and Negotiation* 2013	Identify team knowledge building through communication and comparing processes within and between teams	6 3-person teams of Master's students in Austria participated in an educational simulation that required developing an improvement plan and preparing a report describing it	*Data:* Communication logs from online collaboration interface *Coding:* Manual coding of 12 processes associated with knowledge building *Theoretical Framework:* Fiore et al.'s (2010) macrocognitive processes	*Technique:* Collaboration process analysis (CoPrA) technique *Data Use:* Coded communication strings were used to aggregate frequency counts by 10-minute interval	Plots by team of the frequency of each knowledge building process over time	Teams exhibited different knowledge building patterns by being problem oriented or solution minded, as well as demonstrating varying levels of (1) consensus-oriented behavior, (2) criticism during brainstorming, (3) breadth and/or depth of ideas discussed, and (4) effort to coordinate
Stachowski, Kaplan, & Waller *Journal of Applied Psychology* 2009	Examine relationship between team interactions patterns and effectiveness during crisis situations	14 control room teams, with an average of 4.36 members, from one nuclear power plant in the United States participated in a routine training simulation	*Data:* Video recordings of the 15-minute period following the onset of a crisis *Coding:* Manual coding of 11 behaviors associated with team patterns of interaction *Theoretical Framework:* None	*Technique:* *t*-test comparisons between high and low performing teams *Data Use:* Pattern recognition algorithms embedded in Theme software were used to identify behavioral patterns over time in the coded communication strings; aggregated values representing number of times the pattern occurred, speaker switches, number of members involved in patterns, pattern length, and pattern hierarchy were calculated and used in statistical analysis	None	High performing teams, compared to low performing teams, engaged smaller numbers of speakers in less systematic and less complex interactions that included fewer verbal statements and back-and-forth exchanges
van der Haar, Koeslag-Kreunen, Euwe & Segers *Small Group Research* 2017	Investigate how structuring behavior in emergency management command-and-control teams may benefit constructive conflict and team effectiveness	17 4-6 person multidisciplinary emergency management command-and-control teams participated in a simulation training exercise focused on emergency response	*Data:* Video recordings *Coding:* Manual coding of 10 team leader structuring behaviors (only team leader behaviors are coded) *Theoretical Framework:* None	*Technique:* ANOVA, Mann-Whitney *U* test *Data Use:* Coded communication strings were used to aggregate the frequency counts of behaviors by meeting	None	High performing teams, compared to low performing teams, decrease over time goal orientation, procedural questions, summarizing commands, summarizing decisions, and undirected questions; teams engaging in more constructive conflict decrease over time goal orientation, procedural questions, summarizing commands, summarizing decisions, and directed questions

(continued)

Table 2.1 (continued)

Yoon & Johnson *Educational Technology Research & Development* 2008	Describe group development through communication patterns	7 5-6 member team of Master's students in the United States completed a final course project	*Data:* Chat log transcripts *Coding:* Manual coding of 13 behavioral categories associated with group development *Theoretical Framework:* Wheelan & Hochberger's (1996) group development operation system combined with Gersick's (1988) mid-transition model	*Technique:* Qualitative inductive analysis *Data Use:* Coded communication strings were thematically analyzed	Graphics representing (1) the group development patterns of teams that progressed linearly and teams that progressed adaptively and (2) impeding group development behaviors by team over time	Teams transit through seven group development phases, exhibit one of two distinct group development patterns, and engage in both positive and negative group-development shaping forces
Zoethout, Wesselink, Runhaar & Mulder *Small Group Research* 2017	Examine if the extent to which team members engage in transactivity can explain the emergence of team learning	3 4-8 person vocational education and training teaching teams in the Netherlands engaged in redesigning curricula and/or developing course materials	*Data:* Video recordings *Coding:* Manual coding describing how statements contribute to the conversations, identifying interactional sequences, and coding the type of transactivity represented in the sequences *Theoretical Framework:* Weinberger & Fischer's (2006) modes of transactivity	*Technique:* Qualitative interaction analysis *Data Use:* Coded communication strings were thematically analyzed	None	Team learning does not only focus on taskwork, it also includes goals, responsibilities, work distribution, and member roles; information storage and retrieval occurred more often when the team members were exhibiting transactivity

longitudinal team communication is reported; many of the studies report additional analytical approaches and corresponding results.

Literature Review Trends and Observations

As anticipated the research identified for inclusion in this review represents many fields. The fields were identified by the journals in which the studies were published and include:

- Business (Gerpott et al. 2019; Havakhor and Sabherwal 2018; Hoogeboom and Wilderom 2019; Lehmann-Willenbrock and Chiu 2017; Lehmann-Willenbrock et al. 2015; 2017; Olekalns et al. 2003; Schecter et al. 2017)
- Communication (Lehmann-Willenbrock and Allen 2017; Lehmann-Willenbrock et al. 2013; 2016)
- Computer science (Schneider et al. 2018)
- Education (Lee and Jamison 2003; Rapanta et al. 2013; Yoon and Johnson 2008)
- Engineering (Chiocchio 2007)
- Healthcare (Aufegger et al. 2019; Davis et al. 2016; Gundrosen et al. 2016; Lingard et al. 2002)
- Human factors (Barth et al. 2015; David and Schraagen 2018; Gorman, Cooke et al. 2012; Gorman, Hessler et al. 2012; Gorman et al. 2016; McComb et al. 2010; Miller et al. 2009; Miller et al. 2010; Parush et al. 2014)
- Psychology (Bourbousson et al. 2015; Dunn et al. 2002; Gorman et al. 2010; Gorman and Cooke 2011; Grand et al. 2016; Kauffeld and Meyers 2009; Kennedy and McComb 2014; Kolbe et al. 2014; Lehmann-Willenbrock and Allen 2014; Stachowski et al. 2009)

Several researchers elected to publish in interdisciplinary journals focused specifically on groups and teams (Chiu and Lehmann-Willenbrock 2016; Klonek et al. 2016; Lehmann-Willenbrock et al. 2014; Lehmann-Willenbrock et al. 2011; Seeber et al. 2013; van der Haar et al. 2017; Zoethout et al. 2017).

One of the primary contributions of this review is that it underscores the importance of examining team behaviors, in general, and communication, in particular, longitudinally. Indeed, several of the articles reviewed note that the insights garnered through their research would not have been identified using more static, cross-sectional designs (e.g., Bourbousson et al. 2015; Schecter et al. 2017).

As the purpose of this review is to provide an overview of the various approaches available to longitudinally examine team communication, we summarize research designs; data collection, coding, and manipulation; analytical techniques; visualization techniques; and findings. At the end of this section, additional resources identified through this systematic review of the literature are noted because they provide tutorials for conducting longitudinal research about team communication, and other team behaviors. They were not included in the review because they did not meet the inclusion criteria requiring a theoretical (versus methodological) contribution to team science. Nevertheless, they are useful resources worth noting.

Research Designs

Research subjects represented multiple countries across three continents. Three studies were completed in Australia (Dunn et al. 2002; Miller et al. 2009; Olekalns et al. 2003). In Europe, studies were conducted in Austria (Seeber et al. 2017), France (Bourbousson et al. 2015; David and Schraagen 2018), Germany (Gerpott et al. 2019; Kauffeld and Meyers 2009; Lehmann-Willenbrock and Allen 2014; Lehmann-Willenbrock et al. 2013; 2016; Lehmann-Willenbrock and Chiu 2017; Lehmann-Willenbrock et al. 2017; Lehmann-Willenbrock et al. 2011; 2015; Schneider et al. 2017), the Netherlands (Barth et al. 2015; Hoogeboom and Wilderom 2019; Zoethout et al. 2017), Norway (Gundrosen et al. 2016), Switzerland (Kolbe et al. 2014), and the United Kingdom (Aufegger et al. 2019). North American countries represented were Canada (Chiocchio 2007; Lingard et al. 2002; Parush et al. 2014) and the United States (Chiu and Lehmann-Willenbrock 2016; Davis et al. 2016; Gorman et al. 2010; Gorman and Cooke 2011; Gorman, Cooke, et al. 2012; Gorman, Hessler, et al. 2012; Gorman et al. 2016; Kennedy and McComb 2014; Lee and Jamison 2003; Lehmann-Willenbrock and Allen 2017; McComb et al. 2010; Miller et al. 2010; Stachowski et al. 2017; Yoon and Johnson 2008). Two studies had subjects from multiple countries including the United States and Germany (Lehmann-Willenbrock et al. 2014) or the United States, Bulgaria, Norway, Sweden, and the Netherlands (Schecter et al. 2017). Three studies did not specify the country in which data were collected (Klonek et al. 2016; Rapanta et al. 2013; van der Haar et al. 2017). One study drew their sample from Amazon Mechanical Turk (Havakhor and Sabherwal 2018).

The study samples were drawn from the field, educational classrooms, and laboratories. Participants in the field studies came from a variety of sectors including automotive (Klonek et al. 2016), airline (David and Schraagen 2018), education (Lee and Jamison 2003; Rapanta et al. 2013; Zoethout et al. 2017), emergency response (Dunn et al. 2002; van der Haar et al. 2017), healthcare (Aufegger et al. 2019; Barth et al. 2015; Davis et al. 2016; Gundrosen et al. 2016; Kolbe et al. 2014; Lingard et al. 2002; Miller et al. 2009; Miller et al. 2010; Parush et al. 2014), military (Gorman et al. 2016; Schecter et al. 2017), nuclear (Stachowski et al. 2017), sports (Bourbousson et al. 2015); were drawn from multiple industries within the same studies (Kauffeld and Meyers 2009; Lehmann-Willenbrock and Allen 2014; Lehmann-Willenbrock et al. 2013; Lehmann-Willenbrock et al. 2016; Lehmann-Willenbrock and Chiu 2017; Lehmann-Willenbrock et al. 2017; Lehmann-Willenbrock et al. 2015; Lehmann-Willenbrock et al. 2011); or the researchers did not report the industry (Hoogeboom and Wilderom 2019). Several teams were completing assignments or classroom activities as part of their education including high school students (Chiu and Lehmann-Willenbrock 2016), undergraduate students

(Chiocchio 2007; Schneider et al. 2017), graduate students (Yoon and Johnson 2008), and in one case the authors did not identify the level of students (Gerpott et al. 2019). In the laboratory, subjects were graduate students (Olekalns et al. 2003; Seeber et al. 2013), members of the local public (Gorman et al. 2010; Gorman and Cooke 2011; Gorman, Cooke, et al. 2012; Gorman, Hessler et al. 2012), or undergraduate students (Kennedy and McComb 2014; Lehmann-Willenbrock and Allen 2017; Lehmann-Willenbrock et al. 2014; McComb et al. 2010). Havakhor and Sabherwal (2018) conducted a virtual laboratory-type experiment using members of the public. Grand and colleagues (2016) collected human subjects data, but that aspect of their study was not focused on communication; their work contributing to this review was not derived from human subjects.

The majority of the subjects in the field studies were engaged in their day-to-day activities. Others, however, were participating in simulated training exercises (Aufegger et al. 2019; Davis et al. 2016; Dunn et al. 2002; Gorman et al. 2016; Gundrosen et al. 2016; Schecter et al. 2017; Stachowski et al. 2017; van der Haar et al. 2017).

Data Collection, Coding, and Manipulations

The vast majority of studies used audio and/or video recordings to capture the communication among team members. These recordings were either transcribed and coded (Aufegger et al. 2019; Chiu and Lehmann-Willenbrock 2016; David and Schraagen 2018; Dunn et al. 2002; Gundrosen et al. 2016; Kauffeld and Meyers 2009; Kennedy and McComb 2014; Lee and Jamison 2003; Lehmann-Willenbrock et al. 2016; Lehmann-Willenbrock et al. 2011; McComb et al. 2010; Olekalns et al. 2003; Rapanta et al. 2013) or coded directly (Davis et al. 2016; Gerpott et al. 2019; Hoogeboom and Wilderom 2019; Klonek et al. 2016; Kolbe et al. 2014; Lehmann-Willenbrock and Allen 2014; 2017; Lehmann-Willenbrock et al. 2013; 2014; Lehmann-Willenbrock and Chiu 2017; Lehmann-Willenbrock et al. 2017; Lehmann-Willenbrock et al. 2015; Schneider et al. 2017; Stachowski et al. 2017; van der Haar et al. 2017; Zoethout et al. 2017). Five studies used logs generated by electronic collaboration spaces (Chiocchio 2007; Havakhor and Sabherwal 2018; McComb et al. 2010; Schecter et al. 2017; Seeber et al. 2017; Yoon and Johnson 2008). In seven studies, the researchers coded communication in real time as they observed the subjects collaborating (Barth et al. 2015; Gorman et al. 2010; Gorman and Cooke 2011; Lingard et al. 2002; Miller et al. 2009; Miller et al. 2010; Parush et al. 2014). The data for one study was gathered by asking team members to provide play-by-play analyses of their actions, perceptions, feelings, and thoughts as they watched video of their team collaborating (Bourbousson et al. 2015). In all of these studies, communication content was coded manually by the researchers. Three studies were not manually coded. In two studies, the communication data were automatically generated by the simulation system being used by the researchers (Gorman, Cooke, et al. 2012; Gorman, Hessler, et al. 2012). In one study, communication was transcribed, but coding was automated (Gorman et al. 2016).

The researchers manually coding data were guided by theoretical frameworks in over half of the studies reviewed. The others based their approaches on existing theory, but devised their own approaches (Chiocchio 2007; Chiu and Lehmann-Willenbrock 2016; David and Schraagen 2018; Davis et al. 2016; Dunn et al. 2002; Gerpott et al. 2019; Gorman et al. 2010; Gorman and Cooke 2011; Gorman, Cooke, et al. 2012; Gorman, Hessler, et al. 2012; Gorman et al. 2016; Gundrosen et al. 2016; Kolbe et al. 2014; Lingard et al. 2002; McComb et al. 2010; Parush et al. 2014; Rapanta et al. 2013; Stachowski et al. 2017; van der Haar et al. 2017). The theoretical frameworks included:

- Bales' (1950) interaction process analysis model (Aufegger et al. 2019)
- Bergan and Tombari's (1975) consultation analysis record (Lee and Jamison 2003)
- Fiore et al.'s (2010) macrocognitive processes (Seeber et al. 2017)
- Gersick's (1988) mid-transition model (Yoon and Johnson 2008)
- Hoogeboom and Wilderom's (2015) behavioral categories (Hoogeboom and Wilderom 2019)
- Kauffeld's (2006) Cassel competence grid (Kauffeld and Meyers 2009)
- Kauffeld and Lehmann-Willenbrock's (2012) act4teams coding scheme (Klonek et al. 2016; Lehmann-Willenbrock and Allen 2014; 2017; Lehmann-Willenbrock et al. 2013; 2014; 2016; 2017; Lehmann-Willenbrock et al. 2017; Lehmann-Willenbrock et al. 2015; Lehmann-Willenbrock et al. 2011; Schneider et al. 2017)
- Klein's (2001) phases of team coordination (Miller et al. 2009; Miller et al. 2010)
- Marks et al.'s (2001) recurring phase model (Havakhor and Sabherwal 2018; Kennedy and McComb 2014)
- Miller's (2004) patient information spaces (Miller et al. 2010)
- Pilny et al.'s (2016) relationship event framework (Schecter et al. 2017)
- Pruitt and Carnevale's (1982) social conflict categorization (Olekalns et al. 2003)
- Santos et al.'s (2012) categorisations (Barth et al. 2015)
- Theureau's (2003) course-of-action analysis (Bourbousson et al. 2015)
- Winberger and Fischer's (2006) modes of transactivity (Zoethout et al. 2017)
- Wheelan and Hochberger's (1996) group development operation system (Yoon and Johnson 2008)

The majority of researchers analyzed the coded communication strings either thematically (Gundrosen et al. 2016; Lingard et al. 2002; Yoon and Johnson 2008; Zoethout et al. 2017) or statistically (Aufegger et al. 2019; Chiocchio 2007; Chiu and Lehmann-Willenbrock 2016; David and Schraagen 2018; Davis et al. 2016; Gorman, Cooke, et al. 2012; Gorman, Hessler, et al. 2012; Kauffeld and Meyers 2009; Klonek et al. 2016; Kolbe et al. 2014; Lehmann-Willenbrock and Allen 2017; Lehmann-Willenbrock et al. 2013; Lehmann-Willenbrock et al. 2014; Lehmann-Willenbrock and Chiu 2017; Lehmann-Willenbrock et al. 2017; Lehmann-Willenbrock et al. 2015; Lehmann-Willenbrock et al. 2011; Miller et al. 2009; Miller et al. 2010; Olekalns et al. 2003; Rapanta et al. 2013; Schneider et al. 2017). Some researchers used the communication data in a variety of ways as part of their analyses. These approaches included aggregating data to be statistically analyzed by calculating occurrence frequencies within work-driven time periods such as meetings, surgery phases, or missions (Barth et al. 2015; Gerpott et al. 2019; Gorman et al. 2010; Gorman and Cooke 2011; Gorman et al. 2016; Lee and Jamison 2003; Parush et al. 2014; van der Haar et al. 2017), calculating occurrence frequencies within time-driven periods such as every 5-minute (Dunn et al. 2002) or 10-minute interval (Seeber et al. 2017); and identifying patterns in the communication strings and calculating pattern frequencies at the team (Hoogeboom and Wilderom 2019; Lehmann-Willenbrock and Allen 2014; Stachowski et al. 2017) or individual (Schecter et al. 2017) levels. Other researchers used the communication strings to identify patterns that were converted into temporal strings for analysis (McComb et al. 2010) or used to calculate intersequencial distances between teams (Havakhor and Sabherwal 2018); construct and validate a computational model of team communication (Kennedy and McComb 2014); or qualitatively detect phase breaks (Bourbousson et al. 2015), capture idea longevity (Lehmann-Willenbrock et al. 2016), or interpret results (Barth et al. 2015).

Analytical Techniques

The analytical techniques used to examine team communication were qualitative, computational, and statistical. The qualitative approaches included discourse analysis (Gundrosen et al. 2016; McComb et al. 2010), visual analysis (Lee and Jamison 2003), inductive analysis (Lingard et al. 2002; Yoon and Johnson 2008), and interaction analysis (Zoethout et al. 2017). Two studies were completed using computational modeling and simulation (Grand et al. 2016; Kennedy and McComb 2014), with one also incorporating genetic algorithm optimization (Kennedy and McComb 2014).

The statistical approaches ranged from straightforward statistical comparisons to complex analytical analyses. Statistical comparisons included *t*-tests (Havakhor and Sabherwal 2018; Lehmann-Willenbrock et al. 2016; Stachowski et al. 2017), Mann-Whitney *U* tests (van der Haar et al. 2017), ANOVA (Gorman et al. 2016; van der Haar et al. 2017), repeated measures ANOVA (Gorman and Cooke 2011), and multivariate MANCOVA (Gorman and Cooke 2011). Several researchers employed regression including linear (Lehmann-Willenbrock and Allen 2014), multiple (Gorman, Hessler, et al. 2012), hierarchical (Hoogeboom and Wilderom 2019), time series (Chiocchio 2007; Dunn et al. 2002), and zero inflated Poisson and mixed effects negative binomial (Davis et al. 2016). The other analytical techniques included:

- Collaboration process analysis (Seeber et al. 2017)
- Cross-correlation analysis (Gorman et al. 2016)
- Dynamical analysis (Gorman et al. 2010)
- Event history analysis (McComb et al. 2010)

- Gamma statistical analysis (Olekalns et al. 2003)
- Hierarchical cluster analysis (Gorman et al. 2016; Lehmann-Willenbrock et al. 2016)
- Lag sequential analysis (Kauffeld and Meyers 2009; Klonek et al. 2016; Kolbe et al. 2014; Lehmann-Willenbrock and Allen 2014; Lehmann-Willenbrock and Allen 2017; Lehmann-Willenbrock et al. 2013; Lehmann-Willenbrock et al. 2014; Lehmann-Willenbrock et al. 2015; Lehmann-Willenbrock et al. 2011; Schneider et al. 2017)
- Markov analysis (Miller et al. 2009)
- Multilevel modeling (Gerpott et al. 2019)
- Multiway contingency table analysis (Miller et al. 2010)
- Optimal matching algorithm (Havakhor and Sabherwal 2018)
- Pattern identification (Rapanta et al. 2013)
- Recurrence quantification analysis (Aufegger et al. 2019; Gorman, Cooke, et al. 2012)
- Relational event modeling (David and Schraagen 2018; Schecter et al. 2017)
- Social network analysis (Barth et al. 2015; Bourbousson et al. 2015)
- Statistical discourse analysis (Chiu and Lehmann-Willenbrock 2016; Lehmann-Willenbrock and Chiu 2017; Lehmann-Willenbrock et al. 2017)
- WinPhaser analysis tools (Olekalns et al. 2003)

Interestingly, very little overlap in the analytical approaches exists across research teams. This breadth of analytical approaches underscores how rich these datasets are and the variety of perspectives available to examine longitudinal team communication.

Visualization Techniques

Data visualization was used in numerous articles to highlight various aspects of communication over time. Many of the researchers provided graphical representations of activity over time including frequency counts of the constructs of interest (Chiocchio 2007; Dunn et al. 2002; Gorman et al. 2010; Gorman and Cooke 2011; Lee and Jamison 2003; Miller et al. 2010; Parush et al. 2014; Seeber et al. 2017; Yoon and Johnson 2008) or measures of interest (Bourbousson et al. 2015; David and Schraagen 2018; Gorman, Cooke, et al. 2012; Gorman, Hessler, et al. 2012; Gorman et al. 2016). These plots may include comparisons (e.g., high versus low performing teams (Chiocchio 2007)) or indicators of when team conditions varied (e.g., perturbations were introduced (Gorman et al. 2010)). Running correlations (Bourbousson et al. 2015; Gorman et al. 2016) and overlying kernel density plots (Davis et al. 2016) also depicted dynamic changes over time.

Three articles provided graphical representations highlighting phase transitions. These researchers used recurrence plots (Gorman, Cooke, et al. 2012), phase maps and diagrams (Olekalns et al. 2003), and Markov diagrams depicting transitions between phases for formal and information communication (Miller et al. 2009).

Three articles included visualizations providing summary information about the analytical results. These summaries included social network diagrams across phases (Barth et al. 2015), common lag1 sequences observed in the data (Lehmann-Willenbrock et al. 2013), and a graphical depiction of communication patterns over time (McComb et al. 2010).

Findings

The findings from the articles included in this review offer an array of insights about team communication over time. This summary is at a very high level of detail. Nevertheless, it achieves our purpose in demonstrating the breadth of issues that can be examined through such in depth examinations of longitudinal team communication. The individual papers provide far richer descriptions of study findings and implications for team science.

A variety of outcomes are influenced by teams' patterns of communication including effectiveness (Chiocchio 2007; Hoogeboom and Wilderom 2019; Kolbe et al. 2014; Lehmann-Willenbrock and Allen 2014; Schecter et al. 2017; Stachowski et al. 2017; van der Haar et al. 2017), group mood (Lehmann-Willenbrock et al. 2011), knowledge emergence (Grand et al. 2016), learning (Zoethout et al. 2017), perceived information sharing (Hoogeboom and Wilderom 2019), positivity (Lehmann-Willenbrock et al. 2017), satisfaction (Lehmann-Willenbrock et al. 2013), social support (Aufegger et al. 2019), and viability (Lehmann-Willenbrock and Chiu 2017).

Many researchers examined variations in patterns. For example, research found that patterns change over time (Gerpott et al. 2019; Gorman and Cooke 2011; Kennedy and McComb 2014; Klonek et al. 2016; Lee and Jamison 2003; Olekalns et al. 2003; Yoon and Johnson 2008). Patterns also differ based on work requirements (Barth et al. 2015; David and Schraagen 2018; Davis et al. 2016; Dunn et al. 2002; Gorman et al. 2016; Gundrosen et al. 2016; Miller et al. 2010; Parush et al. 2014), team membership continuity (Gorman et al. 2010; Gorman, Cooke, et al. 2012), member expertise (Bourbousson et al. 2015), country of origin (Lehmann-Willenbrock et al. 2014), team leader behavior (Lehmann-Willenbrock et al. 2015), and whether or not team members show up to scheduled meetings (Lehmann-Willenbrock and Allen 2017).

Finally, comparisons uncovered pattern differences between teams working face-to-face versus via computer interface (McComb et al. 2010); between teams at different performance levels, with different network densities or having different reputation signals (Havakhor and Sabherwal 2018); and between teams comprised of complainers versus solution seekers (Lehmann-Willenbrock et al. 2016) or problem solvers versus solution seekers (Seeber et al. 2017). Patterns are predominately circular or repetitive versus linear (Rapanta et al. 2013), may be enacted by the team to achieve goals, minimize tension, and maintain social cohesion (Lingard et al. 2002), may not progress linearly through phases of coordination (Miller et al. 2009), and are recoverable after perturbations (Gorman, Hessler, et al. 2012).

Specific types of communication may be influential. For example, creativity may be inhibited by rude disagreements (Chiu and Lehmann-Willenbrock 2016), but dysfunctional behaviors can be reduced through humor (Lehmann-Willenbrock and Allen 2014), positive statements (Klonek et al. 2016), or procedural statements (Lehmann-Willenbrock et al. 2013). Moreover, supportive statements may further cycles by reinforcing complaining or encouraging additional conversation about the task assigned to the team (Kauffeld and Meyers 2009) and enhance team-level affect when made after proactive statements (Schneider et al. 2017).

Available Resources

A number of articles omitted during the full paper review phase are worth noting as they provide tutorials about various methods applicable to the examination of team communication over time. For example, they provide detailed instructions for employing recurrence quantification analysis (Knight et al. 2016; Tolston et al. 2019), creating state space grids (Meinecke et al. 2019), developing team process measures (Kalia et al. 2017), conducting auto regressive integrated moving average modeling (Vasileiadou and Vliegenthart 2014), and employing relational event modeling (Pilny et al. 2016). Other researchers offer broader treatises describing a variety of issues to consider and methods available (e.g., Lehmann-Willenbrock and Allen 2018). These tutorials often use data to demonstrate the approaches. They do not, however, draw any conclusions that theoretically inform team science, which was one of our inclusion criteria.

Managerial Minutes

Team communication is understood through science.
Many researchers examine team communication. For example, they have looked for and found patterns in team communication. Through their studies, researchers provide guidance that can be tweaked to fit your particular situation. What's more, the findings may be the seed for a tangential idea that will lead you to your own solutions.

Chapter 3
The Study

The data reported herein were collected as part of a research program, funded by the National Science Foundation and the Office of Naval Research, focused on examining team mental model convergence. Data were collected between 2003 and 2007 at the University of Massachusetts Isenberg School of Management with the approval of the Institutional Review Board.

The Sample

Study participants were recruited from large lecture classes at the junior and senior levels within the Isenberg School of Management. The material taught in these courses was not relevant to the research topic. Students volunteered to participate and were assigned to three-person teams based on schedule availability. All subjects earned course extra credit for participating. Additionally, the teams with the top performance scores were awarded $150 per team, where the odds of winning were 1:5 and the performance comparisons were made within study conditions.

The participants ranged in age from 19 to 46 (M = 21.21 SD = 2.76) and 60% were male. They were juniors (60.30%), seniors (34.61%), or sophomores (5.09%). Finally, the majors represented were 16.28% finance, 13.49% management, 12.47% accounting and 11.96% were sports management. The remaining participants were split among other degree majors (e.g., economics, engineering, marketing, psychology, undeclared) with less than 10% of participants in any one of these majors. The characteristics described aligned with the demographics of the Isenberg School at the time data were collected, thereby indicating a representative sample.

A total of 393 subjects participated, which equates to 131 teams (3 participants per team). Of these 131 teams, 120 teams have data for two performance episodes (T1 and T2) and 11 teams have data for only one episode due to corrupted recording

S. McComb, D. Kennedy, *Computational Methods to Examine Team Communication*, Computational Social Sciences,
https://doi.org/10.1007/978-3-030-36159-4_3

files. We elected to keep these one performance episode teams in the dataset and included in analyses where possible. Our rationale for this decision was to maximize the power available to run the analyses.

The Study Design

All data were collected using a laboratory task without an experimental manipulation. A 2x2x2 between-teams design was used. The study conditions are: performance episode, communication medium, and time availability. All teams completed two performance episodes. Teams were assigned randomly to either a distributed or an in person communication medium. The two time availability conditions were run sequentially, where teams in the unrestricted condition were completed before the time pressure condition started. The laboratory task and study conditions are described below.

Task Description and Procedures

The hidden profile laboratory task used for this research was an intellective task (McGrath 1984) in that it required the subjects to choose a final answer using heuristics and a correct answer can be computed using optimization. Specifically, teams completed a personnel scheduling task, adapted from Bachrach et al. (2001), Earley (1994), and Steele-Johnson et al. (2000), that minimized the amount of prior knowledge brought to it by the team members and maximized the interdependence among them. Each team was tasked with creating one workforce schedule that minimized total cost expressed in employee wages. Workforce requirements (i.e., the number of employees) for each two-hour shift in a seven-day, 12-hour work schedule and a list of 10 employees with different hourly wages were given. A copy of the documents distributed to the team members can be found in Appendix A.

Collaboration among team members was encouraged by randomly assigning team members to specific functional roles. Rules associated with each role provided each team member with unique knowledge about the task requirements and conflicted in such a way that maximizing one role's rules would break the others' rules. The functional roles and associated rules were: (1) human resources representative, each employee can be assigned to work no more than 10 hours per day and 50 hours per week, (2) union representative, each employee must be assigned to work at least 4 hours per day if called into work and 30 hours per week, and (3) operations manager, the schedule requirements must be met exactly. A penalty of 1.8 times the assigned worker's wage was assessed for any broken rules in the final schedule. Subjects were not informed that they each had unique rules.

After volunteering to participate, subjects were informed of the exact time that their team was scheduled. Upon arrival at the laboratory, team members were introduced to one another, seated in a space that was appropriate for their assigned working conditions (described below), and given approximately 5 minutes of verbal orientation about what would occur during the performance episode. Additionally, each team received one-page instruction sheets that were color coded by role, practice schedule worksheets, a final schedule worksheet that was to be submitted at the end of the performance episode, pencils, and calculators. The instruction sheets were identical where the task assignment was explained and unique where the rules associated with the specific roles were described. The instructions sheets for each role can be seen in Fig. A.1 in Appendix A. Also included in Appendix A are the worksheets teams were given for the two performance episodes. Teams could use as many copies as they wanted, but were required to submit a final version.

The teams were allowed to begin after the experimenter ensured that all subjects understood the task. The task was pretested with a comparable sample to ensure that the subjects found the task interesting, the level of difficulty was appropriate for the population, and the task could be completed easily within approximately 60 minutes.

Performance Episodes

Marks and colleagues (2001: 359) define performance episodes as, "distinguishable periods of time … identified by goals and goal achievement." This definition characterizes the work of the teams participating in this research project, as they were given a task with specific goals upon arrival in the laboratory, collaborated to achieve those goals, and then repeated the process for a project with slightly different goals (described below) approximately one week later. The teams, therefore, completed two performance episodes for this research project. The precise time between performance episodes varied slightly because the second performance episode was scheduled at the convenience of the team members.

The rules and employee wages remained the same for both performance episodes but the scheduling requirements changed. In the first performance episode, employee requirements/day were set. In the second performance episode, employee requirements/two-hour time block were given. For example, in the first performance episode three employees were required during every two-hour time block on Days 1 and 2, but five employees were required on Day 3 (see Fig. A.2a in Appendix A). Alternatively, in the second performance episode on Day 1, two employees were required from 8–10 am and 10–12 am, but 3 employees were required for the remaining two-hour time blocks that day (see Fig. A.2b in Appendix A).

No feedback was given to the teams about their performance after either performance episode. We made the decision not to give feedback because of the positive relationship between feedback and performance (Erez 1977). By not giving performance feedback, we are able to isolate the effects of team communication.

Communication Media

Teams were randomly assigned to one of two communication media study conditions: distributed or in person. Both conditions required synchronous communication, but the communication media differed. In person teams were placed in a private room where they could complete the task via face-to-face communication. A digital timer was used to record elapsed time. Team communication was recorded and transcribed verbatim. Transcripts were reviewed by a second person to verify accuracy.

Distributed team members were assigned to one of 10 personal computers located in a research laboratory and linked in a local area network. Seat assignments were made such that no team members were sitting next to each other. This arrangement ensured that the team members could not see each other's notes or engage in oral communication even though they were physically located in the same laboratory. The only communication allowed among team members was via a synchronous computer conferencing system specifically developed for this laboratory task. The system consisted of a screen with the task requirements that was identical to the paper copy given to all team members, a text messaging area, and a timer depicting the elapsed time since the performance episode started. Each team had its own unique collaboration environment where they could work synchronously to create the workforce schedule. All team members had equal access to the screen and could assign/delete workers, view any assignments or deletions made by teammates, and communicate with the team. The postings in the text box were identified with the sender's name and appeared in the order in which they were sent. The team members could scroll through their communication history. In addition to the aforementioned five-minute orientation, distributed teams were given a 10-minute orientation to the collaboration environment to ensure they understood its functionality. Entire performance episodes were recorded and the communication histories served as transcript data for this research.

Teams completed both performance episodes using the same communication medium. For the larger study, a mixed communication medium condition was used where teams completed the first performance episode in person and the second performance episode via the distributed condition. Two publications based on data from the larger research study focused only on in person teams completing performance episode one (Kennedy and McComb 2014; McComb et al. 2014). The mixed media teams were included in those studies to increase the sample size. Since some results reported in this book are longitudinal, these mixed media teams are not included in any of the analyses.

Time Availability

The time availability study conditions were unrestricted time and time pressured. Teams assigned to the unrestricted time condition were given unlimited time to complete the task and were unbothered throughout the performance episode. In other words, teams were allowed to stay until the team members decided that they were satisfied with their schedules. Under time pressure, teams were given 45 and 30 minutes to complete their schedules in the first and second performance episodes, respectively. These time limits were established based on the research team's experience with the unrestricted time condition to allow the majority of teams adequate time to finish if they maintained their focus on the task.

The Data

Summary data by condition are provided in Table 3.1, including sample sizes, performance metrics, and communication string lengths. Specific descriptions of how performance metrics and communication string lengths were obtained follow.

Performance Measures

Straus and McGrath (1994) suggest that quantity, quality, and speed are the primary indicators for assessing team performance. Given that the task required completion of one schedule, quantity is irrelevant for our purposes. Performance quality was determined by the employee cost associated with the final schedule submitted by each team. Performance speed was represented as the time taken to achieve that

Table 3.1 Sample sizes, time and cost performance, and communication string length by study condition

Time condition	Media	Performance episode	Sample size	Cost *M*(*SD*)	Time *M*(*SD*)	Communication string length *M*(*SD*)
Unrestricted	Distributed	T1	35	3070.42 (277.97)	84.51 (27.47)	273.46 (123.29)
		T2	36	2519.19 (138.08)	53.69 (16.76)	201.72 (98.09)
	In person	T1	35	2964.79 (164.62)	65.77 (23.78)	299.40 (207.77)
		T2	29	2470.66 (60.01)	41.31 (12.19)	190.55 (149.38)
Time pressured	Distributed	T1	30	3549.16 (528.18)	45.00 (0.00)	146.27 (68.23)
		T2	30	2884.54 (562.46)	29.63 (2.01)	119.00 (43.14)
	In person	T1	28	3610.16 (825.32)	44.39 (3.12)	419.23 (179.46)
		T2	28	2705.00 (463.52)	28.84 (3.26)	305.89 (143.27)

cost. The cost and time performance achieved by the teams are summarized in Table 3.1.

The work schedule cost represents performance quality and was computed by (1) adding up the number of hours each employee worked, multiplying it by the employee's wage, and summing the cost for all employees and (2) checking the schedules to determine if any rules were broken and assessing a penalty of 1.8 times the respective employee's wage for any infractions. This metric resulted in a performance measure where smaller values represent superior performance. Given the conventional norm of "more is better" and to aid in interpretation, these values were transformed during analysis. The optimal costs for the schedules in the first and second performance episodes were calculated. A team's quality performance score was transformed by dividing the optimal cost by the team's score. As such, each team's cost was divided by $2438 and $2430, for the first and second performance episodes, respectively. The resulting scores range from 0 to 1, where 1 represents the case where a team achieved the optimal score.

The time required for teams to complete the task were recorded to represent performance speed. For the time pressure condition, teams were allotted 45 minutes to complete the first performance episode, and 30 minutes to complete the second performance episode. Interestingly, some of the time pressure teams (1 distributed team at the first episode, 5 distributed teams at the second episode, 2 in person teams at the first episode, and 5 in person teams at the second episode) finished before their allotted time was up. Because smaller values of this metric also represent better performance, it was transformed by subtracting a team's recorded time from the slowest team's time plus one. For the first performance episode, the slowest team took 162 minutes, and in the second, the slowest team took 160 minutes. Using the conversion, the slowest team has the lowest score of one and the fastest team has highest score, which represents how many minutes faster they completed the task than the slowest team.

Communication Strings

The transcripts of communicated messages from distributed teams were captured by the synchronous computer conferencing system through which the team members collaborated via the teams' texted communication histories. The recordings of in person team communication were transcribed into a string of messages that occurred sequentially over time. The resulting transcripts represent message-to-message communication strings that could be processed by data coders.

To conduct our longitudinal analyses examining team communication, the distributed communication histories and in person transcripts were transformed into coded communication strings. The coding framework and process used to convert the transcripts to coded communication strings are described below.

Coding Framework

Originally, an inductive approach was used to devise a coding framework whereby two researchers independently examined each message and iteratively devised a coding framework that represented the majority of content embedded in the contextual language (Bohm 2004). Using this approach, a two-dimensional coding framework was devised that represented both what the teams were discussing (i.e., the topics being discussed) and why they were delivering specific messages (i.e., the purpose). This original coding framework for topic and purpose is reported in McComb et al. (2014) and for topic only in Martínez Ayala et al. (2017). More detail about similarities and differences between these studies and the research reported herein is provided at the end of this chapter.

The topic codes aligned with Marks et al. (2001) transition processes (i.e., goal specification, mission analysis, strategy formulation) and one general action process. We, therefore, applied this theoretically grounded coding framework. During the alignment process, we observed that strategy formulation entailed both discussing tactical strategies for achieving the task (e.g., “do we want to choose who will work on what days?”) and operational strategies regarding members’ roles and responsibilities (e.g., “I’ll assign the workers while you calculate the hours”). Moreover, these two distinct types of strategy formulation were often discussed by team members in separate conversations, thereby suggesting that they may be two unique processes worthy of investigating independently. In Kennedy and McComb (2014, p. 788–9), we described the archival support for separating them into two unique process codes as follows:

> “The strategic formulation process sets the approach for teams to achieve their objectives (Marks et al. 2001). This preparation engages members in planning that helps them coordinate various activities (Mehta, Feild, Armenakis, and Mehta, 2009). Notably, researchers have indicated that making a strategic plan improves task performance (Hackman and Morris 1975; Janicik and Bartel 2003; Weingart 1992) and is particularly important early in taskwork (Hackman et al. 1976; Woolley 1998) before action processes are initiated (DeChurch and Haas 2008; Katzenbach and Smith 1999). Yet researchers have suggested that planning is composed of different dimensions (Stout, Cannon-Bowers, Salas, and Milanovich 1999). In particular, planning may involve what, when, and how actions will be taken as well as who will be responsible for certain actions (Stout et al. 1999). Indeed, the generation of members’ roles and responsibilities may be a separate type of planning activity that is pertinent to team effectiveness (Cannon-Bowers, Salas, and Converse 1993; Salas et al. 1999) and may occur as its own process during team development (Kozlowski et al. 1999). Mathieu and Rapp (2009) demonstrated the importance of demarcating different types of plans. In their study, a tactical plan formalized the team’s strategy performance, and an operational plan, in the form of a team charter, established members’ roles and responsibilities.”

Therefore, we elected to separate strategy formulation into tactical strategy and operational strategy.

The resulting process codes, content definitions, and example messages are listed below. This coding framework was previously applied for the research reported in Kennedy and McComb (2014) and Knight et al. (2016). The topic codes, including definitions and example messages are:

- *Mission analysis* is assigned to messages about the task objectives, environment, and resources. Example: "how many hours can workers work per week?"
- *Goal specification* is assigned to messages about task goals and the prioritization of goals for task accomplishment. Example: "let's win the $50."
- *Tactical strategy* is assigned to messages about how the team will approach the assigned task. Example: "we should assign workers with lower wages first."
- *Operational strategy* is assigned to messages about the roles and responsibilities of members. Example: "I'll be responsible for keeping track of employee hours worked."
- *Action* is assigned to messages about actual performance of the assigned task. Example: "I put Bob on Day 2 from 4 to 6."

Whereas the topic codes capture what the team is conveying, the purpose codes capture why the team member is sending the message (McComb et al. 2014). No theoretical foundation was identified for the purpose coding framework. Thus, we have conducted our analyses using the inductively generated codes that represent why teams are exchanging messages about particular topics. The result is a set of seven purpose codes. The purpose codes, including definitions and examples, are:

- *Action* is assigned to messages describing actual activity being done by a team member. Example: "Putting Al on Day 3."
- *Agree* is assigned to messages either agreeing or disagreeing. Example: "right."
- *Apply* is assigned to messages about how the topic is being used by the team to accomplish the task. Example: a *tactical strategy-apply* statement was, "ok, we can put the cheap people in 25 times each … making sure not to go over 10 hours per day."
- *Clarify* is assigned to messages that ask a question. Example: "can employees work more than one shift in the day?"
- *Inform* is assigned to messages about facts or task goals. Example: "just no more than 10 hours per day."
- *Suggest* is assigned to messages that are recommendations. Example: "but we should stick to [Al's] because it's less."
- *Summarize* is assigned to messages summing up what the team has already done or shared. Example: "so less than ten and at least four [hours]."

Coding Process

Coding was completed by two researchers unfamiliar with the research purpose. A coding example from a partial transcript for an in person unrestricted team in their first performance episode is provided in Table 3.2. The process started by training

Table 3.2 Example coded transcript

Message Number	Topic	Purpose	Message
1	Tactical strategy	Inform	We're giving the least expensive the most hours and the most expensive the least hours
2	Tactical strategy	Inform	We gave everyone the minimum and this is what we have left
3	Tactical strategy	Inform	Oh I see
4	Tactical strategy	Apply	Now we have the hours they should and we need to make sure nobody works less than four or more than ten hours
5	Tactical strategy	Suggest	Per day
6	Tactical strategy	Clarify	Yeah do you want to start with the mostexpensive or the least expensive
7	Tactical strategy	Clarify	You want to do the most expensive or do you want to do the least expensive
8	Tactical strategy	Suggest	We'll do the least the expensive we can fill up the most hours put them ten hours per day
9	Tactical strategy	Suggest	So A B C just do like day 1 day 2
10	Tactical strategy	Inform	Our least expensive is going to work fifty hours so give them five ten hours shifts
11	Operational strategy	Suggest	Okay you want to do that
12	Tactical strategy	Suggest	We don't have to do that we can make it something else
13	Tactical strategy	Suggest	Well let's do that we don't know how it's going to end up
14	Tactical strategy	Suggest	Actually let's start the other way we could put the least hours in first and leave a lot of hours so they can fill it in well I guess it doesn't matter
15	Tactical strategy	Suggest	We can have like five people work on this day
16	Tactical strategy	Apply	That's what's required
17	Tactical strategy	Inform	Which is three each hour
18	Tactical strategy	Suggest	No I think its three the whole day
19	Tactical strategy	Suggest	I think we could have three different people in each one
20	Tactical strategy	Suggest	You think so I think its three no more than three names here not at one time for the whole day
21	Tactical strategy	Clarify	You think so
22	Tactical strategy	Suggest	If the schedule calls for two people let's make sure though you want to just confirm it, okay 3 people per box
23	Tactical strategy	Suggest	3 people per box that's what we calculated for right, so let's take our cheapest guys first
24	Tactical strategy	Agree	Yeah let's do the three cheapest
25	Action process	Inform	A B and C
26	Action process	Action	Give them ten hours each ten right there you know what we can't that leaves two hours and no one can work two hours a day
27	Action process	Apply	They can't only work two shifts okay
28	Action process	Inform	Three boxes is good
29	Action process	Inform	Six hours
30	Action process	Suggest	No 8

(continued)

Table 3.2 (continued)

31	Action process	Agree	Yeah 8 hours
32	Action process	Suggest	8 times what is fifty
33	Action process	Inform	6.25
34	Action process	Suggest	What if we did 6, 50 divided by six 8.33, we can just count it we don't have to do it like that give them four 8-hour shifts
35	Action process	Inform	That's 32
36	Action Process	Action	Give them 5, we can't have 48
37	Action process	Suggest	We can still add people to this shift
38	Action process	Suggest	How many do we have so far 5 times 8 that's forty hours ten more hours put them on a ten-hour shift on this day
39	Action process	Action	So ten hours
40	Action process	Inform	That's it
41	Action process	Suggest	That's it waitwe can't give them ten hours
42	Action process	Inform	That's fifty
43	Action process	Inform	So it's good
44	Action process	Inform	Yeah so 8 times 5 is forty plus ten is fifty
45	Action process	Inform	So we used them all at the maximum
46	Operational strategy	Inform	Okay so I'm going to check these guys off they're checked off so now we'll go to the next ones the 42 hours alright B and D
47	Operational strategy	Inform	I'm going to break this down just because we have space
48	Action process	Inform	Weneed one more person here two more workers and four more workers so then for me this makes sense
49	Action process	Suggest	Do you want to finish these days off these fourhour shifts give our most expensive guys these four hours shifts let's fill up their hours so they have to work thirty hours right
50	Action process	Suggest	No 42
51	Action process	Suggest	No the most expensive
52	Action process	Clarify	Why are you going to them let's just go down the thing
53	Action process	Suggest	You want to do that we can do that

the coders, which entailed ensuring they understood the codes and definitions, how to apply them to the statements in the transcripts, and the importance of frequent comparisons. The coders and primary researchers then worked together to code one transcript. Next, the two coders independently coded one transcript, reviewed it with the primary researchers, and discussed any outstanding questions they had regarding the codes and coding process. This iterative process continued until both coders were comfortable. The coders then met periodically to compare their coded transcripts. When the coders disagreed about the codes for a particular message, they would discuss the message, their logic for assigning a particular code, and come to an agreement about what code to assign.

As the coding process unfolded, the coders found individuals' statements that contained multiple topics and/or purposes. These statements were then parsed into messages representing only one topic and purpose. The result of this process was a transcript containing individual coded messages comprised of exclusive information (Chi 1997) that could be transformed into communication strings. The lengths of the communication strings (i.e., the number of messages) by condition are given in Table 3.1.

Meyers and colleagues (1991) suggest that coding schemes are adequate when coders can demonstrate that they (1) possess a general understanding of the codes and corresponding definitions, and (2) agree upon how the codes should be applied consistently. The two coders achieved 87%–99% agreement per transcript, thereby supporting the adequacy of the two-dimensional coding framework.

Data Operationalizations

For the analyses reported in this book, the data were operationalized in different forms. For the analyses reported in Chapters 5 and 7, the communication strings were divided into quartiles (Q1-Q4). The motivation for separating the communication strings into time segments was based on previous research suggesting that team discussions may vary over time (e.g., Hewes and Poole 2012; Kennedy and McComb 2014; Pavitt 1992; Pavitt and Johnson 2001) and smaller units may be required to capture nuances in these variances (Ballard et al. 2008). Finally, Gersick (1988, 1989) identified that a critical shift in team processes occurs at a team's midpoint. We parsed the data into quartiles to capture any nuances that may occur within team activities before and after the midpoint. The communication strings were broken into quartiles based on communication frequency (versus elapsed time) as Patrashkova and McComb (2004) found no statistically significant differences between communication frequency and duration results.

In Chap. 4, to study the rhythm of team communication, we use the communication string depicting the order in which topics were discussed. The analyses also include team performance.

In Chap. 5, we use topic frequencies and $t + 1$ transitions from the communication strings to examine team communication timing and pacing, respectively. The topic frequencies are the tallies of how many times each topic was discussed by the team. These frequencies were calculated for each quartile and used to examine team communication timing. For example, in the transcript excerpt provided in Table 3.2, the topic frequencies for action, goal specification, mission analysis, operational strategy, and tactical strategy are 27, 0, 0, 3, and 23, respectively. The $t + 1$ transitions capture conversation progression from one message to the next. Using the partial transcript provided in Table 3.2, the first $t + 1$ transition is from tactical strategy to tactical strategy; later in the transcript, the $t + 1$ transition between the tenth and 11th messages is tactical strategy to operational strategy. The frequencies of these $t + 1$ transitions by quartile were calculated and used to examine team communication pacing.

In Chap. 6, points where conversations shift among topics are employed to examine the polyphony associated with team communication. The process used to identify conversation shift points was developed for the research reported in Kennedy and McComb (2014) and is described here. The shift point for the topic of action is noted to be the first time teams shift to action in the communication string; therefore it is recorded as the message position where the first action message occurs. For the sample transcript provided in Table 3.2, the shift point to action occurred at Message 25.

The shift points about transition process topics were devised to represent when teams ended meaningful conversations (at least three messages about the same topic) and shift to other topics for an extended period of time (more than 25 messages). Only one transition process topic shift point occurs in the sample transcript provided in Table 3.2; it occurs at Message 24 when the last tactical strategy message is exchanged before the team shifts to other topics for more than 25 messages. Team performance is also included in these analyses.

The rule about the number of messages before and after a transition process shift was constructed in two phases. First, a subject matter expert reviewed 10% of the transcripts to determine (1) how many messages may be needed to be considered a substantive conversation about a transition process and (2) how many messages had to be exchanged before revisiting that topic could be considered a new conversation (versus an extension of the original conversation). The results were definitive for needing at least three messages to have a meaningful topic exchange. Less clear was the number of messages required for the next communication about that topic to be considered a new conversation. Therefore, mathematical analysis was undertaken to ensure that the cutoff criterion captured the occasions identified by the subject matter expert as demonstrating a completed conversation and ensuring that no completed conversations were nullified by the rules established. The results of this investigation indicate that 25 messages are the most appropriate cutoff criterion.

In Chap. 7, we return to examining $t + 1$ transitions by quartile, but we include both the topic and the purpose to examine sequences in team communication.

Previous Data Use

The data used for the analyses reported in this book are only a portion of the data collected as part of a much larger research program focused on examining team mental model convergence. For example, questionnaire data were also collected but are not relevant to this examination of longitudinal team communication. Other study conditions were also included in the larger effort, specifically a third communication media condition was included, where teams met in person for the first performance episode and distributed in the second performance episode, and a third time availability condition was included, where teams were interrupted and given new information after a period of time had elapsed. These study conditions were omitted to streamline our investigation of team communication and ensure a concise focus that can be clearly explicated to our readers.

Nevertheless, portions of the communication string dataset used for this research study have been previously published. We provide the following brief descriptions of those publications to assure our readers that the research published herein represents a unique contribution to the literature about longitudinal team communication.

First, Kennedy and McComb (2014) is a theoretically-driven examination of team communication, but the methodological approach is computational modeling, simulation, and optimization. Topic only communication strings from unrestricted in person teams' first performance episode were combined with the data from the unrestricted teams in the mixed media condition who met in person for the first performance episode. The data were used to validate the computational model and as a baseline comparison for the results of virtual experiments conducted by simulating various scenarios with the computational model and the optimization procedures.

Second, Knight et al. (2016) use topic communication strings from unrestricted teams, both distributed and in person, to demonstrate how recurrence quantification analysis (RQA) can be used. We employ RQA as part of our investigation of team communication reported herein on a larger corpus of data. Moreover, the foci of the article and this investigation are different. In the article, the impetus is on demonstrating how to conduct RQA and the data are used to demonstrate the methodological techniques; no implications for team communication are drawn. We include some instruction about how to deploy the method in the chapter appendix. However, our primary emphasis in this book is to draw inferences from the RQA results that contribute to the team communication literature.

Third, McComb et al. (2014) is a very short conference paper that explores the viability of employing hierarchical clustering with limited theoretical development. The preliminary analyses use a subset of data (i.e., topic and purpose communication strings from unrestricted in person teams during the first performance episode) and an early version of the communication coding framework that is not theoretically grounded (as the framework is now). Thus, this paper has the most overlap with what is reported herein, but the scope is far more limited, and less theoretically robust, than our current effort.

Finally, Martínez Ayala et al. (2017) is a methodological treatise focused on various approaches to data abstraction. Topic and purpose communication strings from unrestricted in person teams during the first performance episode were used as a test case for comparing the abstraction approaches. Thus, the methodological approaches and findings have no overlap with the research presented in this book.

Managerial Minutes

Find time to make simple adjustments to team communication.
In this chapter, we focus on the data collected for our research. Improving team performance in the field, however, does not always require a formalized, data-driven examination. Sometimes simple adjustments can result in observable differences. These adjustments can range from engaging the team in a conversation about an ongoing issue with communication or collaboration to intervening in a specific way. Nevertheless, identify a few appropriate measures so you can monitor the impact of your efforts.

Be ready to make adjustments to team communication.
If an intervention seems like the best path forward, start by identifying the outcome you want to impact (e.g., team cohesion). Next, consider various options for influencing it (e.g., training about task versus relationship conflict) and try something (e.g., have a conversation about conflict). Pay attention to any changes in how the team communicates, particularly with respect to the outcomes you are hoping to impact.

Appendix: Example Dataset Description

One of our goals with this book is to provide detailed, accessible instructions describing how we conducted the various analyses. These instructions will be located at the end of each chapter in a chapter appendix. For this chapter, we introduce the example dataset we used throughout these chapter appendices.

Data Description

To facilitate demonstrating how to conduct the various analyses presented herein, we have simulated an example dataset with properties that are similar to those of the dataset used in this research study. The dataset is presented in Table 3.3. First, this dataset includes coded communication strings for 10 teams. Second, teams are assigned to one of two conditions that may influence communication. Third, a scaled performance score ranging from 0 to 1 is provided. Finally, the communication strings are comprised of a two-digit number to make them analogous to our topic and purpose codes. In this example dataset, we have included four topic and three purpose codes. Since some analyses are conducted using only the topic data, the strings from the top half of Table 3.3 were transformed into strings with only topic information and listed in the bottom half of Table 3.3.

Table 3.3 Example dataset used throughout chapter appendices

Team	Condition	Performance	Communication string (topic and purpose)
Team 1	1	0.98	11, 12, 11, 21, 23, 23, 23, 31, 32, 31, 12, 13, 12, 41, 41, 42, 41, 41, 41, 41
Team 2	1	0.85	11, 23, 22, 23, 21, 33, 11, 12, 12, 11, 31, 32, 31, 31, 22, 22, 23, 41, 42, 41, 41, 42
Team 3	1	0.75	22, 11, 22, 23, 22, 32, 31, 12, 13, 11, 12, 11, 31, 32, 31, 11, 12, 11, 42, 41, 42, 41, 41, 41
Team 4	1	0.90	12, 11, 12, 31, 32, 22, 23, 23, 22, 23, 31, 32, 31, 41, 11, 11, 12, 41, 42, 42, 42, 42
Team 5	1	0.40	22, 21, 22, 22, 11, 12, 31, 32, 12, 11, 12, 11, 42, 41, 32, 31, 32, 41, 22, 41, 42, 41, 41, 41
Team 6	2	0.65	31, 32, 31, 12, 13, 12, 11, 42, 23, 22, 21, 32, 32, 31, 12, 12, 11, 41, 42, 42, 42, 42
Team 7	2	0.30	11, 12, 12, 12, 31, 32, 31, 41, 11, 12, 11, 11, 12, 11, 41, 22, 21, 41, 42, 42
Team 8	2	0.40	41, 32, 31, 32, 12, 11, 11, 22, 22, 22, 31, 32, 32, 32, 12, 12, 11, 41, 42, 42, 41, 41
Team 9	2	0.22	11, 12, 22, 42, 41, 11, 12, 11, 32, 31, 32, 42, 22, 23, 22, 42, 41, 42, 11, 12, 11, 11
Team 10	2	0.30	12, 13, 11, 23, 22, 22, 31, 32, 33, 12, 13, 11, 23, 22, 21, 42, 41, 11, 12, 13, 41, 42
			Communication string (topic only)
Team 1	1	0.98	10, 10, 10, 20, 20, 20, 20, 30, 30, 30, 10, 10, 10, 40, 40, 40, 40, 40, 40, 40
Team 2	1	0.85	10, 20, 20, 20, 20, 30, 10, 10, 10, 10, 30, 30, 30, 30, 20, 20, 20, 40, 40, 40, 40, 40
Team 3	1	0.75	20, 10, 20, 20, 20, 30, 30, 10, 10, 10, 10, 10, 30, 30, 30, 10, 10, 10, 40, 40, 40, 40, 40, 40
Team 4	1	0.90	10, 10, 10, 30, 30, 20, 20, 20, 20, 20, 30, 30, 30, 40, 10, 10, 10, 40, 40, 40, 40, 40
Team 5	1	0.40	20, 20, 20, 20, 10, 10, 30, 30, 10, 10, 10, 10, 40, 40, 30, 30, 30, 40, 20, 40, 40, 40, 40, 40
Team 6	2	0.65	30, 30, 30, 10, 10, 10, 10, 40, 20, 20, 20, 30, 30, 30, 10, 10, 10, 40, 40, 40, 40, 40
Team 7	2	0.30	10, 10, 10, 10, 30, 30, 30, 40, 10, 10, 10, 10, 10, 10, 40, 20, 20, 40, 40, 40
Team 8	2	0.40	40, 30, 30, 30, 10, 10, 10, 20, 20, 20, 30, 30, 30, 30, 10, 10, 10, 40, 40, 40, 40, 40
Team 9	2	0.22	10, 10, 20, 40, 40, 10, 10, 10, 30, 30, 30, 40, 20, 20, 20, 40, 40, 40, 10, 10, 10, 10
Team 10	2	0.30	10, 10, 10, 20, 20, 20, 30, 30, 30, 10, 10, 10, 20, 20, 20, 40, 40, 10, 10, 10, 40, 40

Each appendix will include two sections: (1) information about preparing the data to conduct the analysis being presented and (2) instructions for how to conduct the analyses, including results for the example data. The results are included for readers interested in trying the analytical approaches by replicating the analyses using the example data.

Chapter 4
Rhythm of Team Communication

Team conversations ebb and flow as members collaborate to complete their tasks. This ebb and flow may be characterized as having varying degrees of stability, predictability, and complexity. Team leaders are in a position to facilitate conversations that help keep team members systematically working their way through the transition and action processes needed to accomplish their work. In this chapter, our efforts to understand the rhythm of communication will inform how team leaders can best guide teams through their conversations. Our results will underscore when teams may benefit from focused versus cyclic conversations, whether those conversations should follow predictable patterns or be seemingly random, and if the time spent talking about specific topics should be consistent or vary over time. These insights may be dependent upon a variety of issues including where in the lifecycle the teams are, how much time they have to complete their task, and what media is available for communicating. Characterizing the rhythm of communication among team members collaborating under various working conditions will be useful in providing team leaders with clear expectations about how conversations may flow over time and when interventions may be needed to help teams get back on track.

Introduction to Team Communication Rhythm

In this chapter, we examine the rhythm of team communication using recurrence quantification analysis. The rhythm of organizational change is revealed through the repetition of cycles, the shifts between focused and switching activities, and the way the activities slow down or speed up over time (Bartunek and Woodman 2015). These same patterns are observable in team communication as team members converse about transition processes (i.e., goal specification, mission analysis, operational strategy, tactical strategy) or the action they are undertaking. Thus, we

S. McComb, D. Kennedy, *Computational Methods to Examine Team Communication*, Computational Social Sciences,
https://doi.org/10.1007/978-3-030-36159-4_4

examine how the rhythm of team communication unfolds over time, and if it differs between high and low performing teams.

A number of theories about teamwork may inform the level of nuance needed to explore rhythm as it relates to team activities across time. For example, punctuated equilibrium by Gersick (1988, 1989, 1991) proposes a phase transition of teamwork around the temporal midpoint, where teams may attempt different approaches to taskwork before the midpoint and shift to more constructive action after the midpoint. As well, entrainment by Ancona and Chong (1999) suggests that teams will attune their activities to external events over time (e.g., budgetary cycles). Such accommodation may result in varying patterns and cycles of communication as team's adjust their activities to temporally coordinate with the external impetus. Simultaneously, Mitchell and James (2001) propose that time lags are critical to the nature of organizations, such that more needs to be known about the lag between when one event occurs and when another event or effect follows. Capturing team conversation lags between, for example, focused conversations or periods of switching among topics may be informative. To capture these rhythmic phenomena, we employ recurrence quantification analysis as it facilitates examining patterns, both visually and statistically, within team communication data across time.

Recurrence quantification analysis (RQA) is an approach developed by systems dynamics scholars to study the patterns of change in non-linear systems (Eckmann et al. 1987). RQA began as a graphical approach called a recurrence plot that could be used to visualize recurrent activity by the system (Marwan 2008). Over time, researchers generated recurrence quantification metrics that facilitate comparisons across systems of interest (Eckmann et al. 1987). The metrics most applicable to our examination of longitudinal team communication are the recurrence rate, determinism, and entropy of the system, which represent a system's stability, predictability, and complexity, respectively.

To date only a paucity of research applying RQA has used groups of more than two members (e.g., Fusaroli et al. 2016; Gorman et al. 2012). Yet, this approach provides nuanced information about the progression of team activities that may help researchers explore longitudinal research questions. For instance, Strang and colleagues (2014) focus on the physio-behavioral coordination of dyad members playing a video game by processing the postural sway data and cardiac interbeat intervals to plot and assess recurrence. The results show that coupling was more likely than by chance; however, higher coupling was negatively related to performance and other team attributes like cohesion and trust. The results provide insights about complementary coordination that occurred and the way dyad member roles may affect coordination. As can be seen in the results of Strang and colleagues (2014), RQA has great potential to illuminate cycles of interconnected activity. This potential is further explicated in the tutorial prepared by Knight and colleagues (2016), where they bridge the divide between the discussion of recurrence analysis within the systems dynamics literature and the kinds of research questions that group researchers study, thereby illustrating how the technique can be applied to teams systems with more than two people.

In our examination of team communication, we anticipate that differences in rhythm will be observed over time. As Gersick (1988, 1989, 1991) observed, teams collaborate differently before and after the midpoint of their life cycle, which suggests that they may require specific communication rhythms to facilitate these differences. More recently, other scholars have also identified changing patterns of communication over time (e.g., Gerpott et al. 2019; Gorman and Cooke 2011; Kennedy and McComb 2014; Klonek et al. 2016; Lee and Jamison 2003; Olekalns et al. 2003; Yoon and Johnson 2008).

As discussed in Chap. 1, normal collaborative processes may be disrupted under time pressure (David and Schraagen 2018; Davis et al. 2016; Driskell et al. 1999; Ellis 2006; Kelly and McGrath 1985) and differences in information sharing capability may result from the different communication media available to the teams in this study (Daft and Lengel 1986). Thus, we also anticipate differences in the rhythm of communication among teams collaborating via different working conditions (i.e., time availability and communication media). Finally, seemingly similar teams often achieve varying levels of team performance. These performance differences may be attributable to the teams' abilities to effectively collaborate and progress toward their goals (DeMatteo et al. 1998; Salas et al. 2018). This collaboration may be facilitated by the manner in which the teams communicate, thereby suggesting that differences in the rhythm of communication may result in differences in team performance.

Research Questions

The research questions associated with rhythm are:

RQ4.1. Do teams have a rhythm in their communication over time?

RQ4.2. Does the rhythm of communication differ depending on the time they have available (i.e., unrestricted time versus time pressured)?

RQ4.3. Does the rhythm of communication differ depending on the communication media available to them (i.e., distributed versus in person)?

RQ4.4. Does the rhythm of communication differ depending on the cost performance achieved?

Methodology

Methodological Approaches: Recurrence Quantification Analysis and Linear Mixed Modeling

The recurrence quantification analysis (RQA) approach was selected to examine team communication rhythm because its results offer insights into the patterns of communication over time in three specific ways. First, the recurrence plots provide a graphical representation of the data that can be used to visually compare how teams differ across the performance episode in terms of how often teams change topics or stay on a topic as the members collaborate. From the visualizations we gain a perspective of the rhythm of conversations and answer our first research question. Second, RQA includes metrics that can be calculated after translating the graphical representations into numerical values. As indicated above, the metrics relevant to our investigation are the recurrence rate, determinism, and entropy (note that additional metrics are available in Marwan and colleagues' (2007) comprehensive review of recurrence analysis).

Third, statistical analyses can be conducted using the RQA metrics. Because the data are captured over two performance episodes, we use a linear mixed model with repeated measures approach. In order to appropriately control for minutes of time taken to achieve the task where variability matters, we run the regression analyses separately for each condition. The results provide the evidence needed to answer RQ4.2, RQ4.3, and RQ4.4.

Data Specification

The communication strings comprised of the topics discussed by each team for each episode were assessed separately to create the recurrence plots and to ascertain the RQA metrics. The raw data are ordered vectors representing teams' changing states over time and, therefore, can be transformed directly into a recurrence plot. The data can also be transformed into matrices of zeros and ones for use in calculating the RQA metrics.

Each of the RQA metrics have different input requirements. Specifically, the recurrence rates are derived from the matrices of zeros and ones. Similarly, determinism is calculated directly from the matrices of zeros and ones where straight diagonal lines of ones are found, but the calculation also requires defining a lower threshold of how many ones in a line are required to represent when teams have demonstrated consistency over time (Marwin et al. 2007). For our data, this consistency can be operationalized as when teams have engaged in a meaningful conversation about a topic. As described in Chap. 3, we inductively ascertained that at least three consecutive messages about a single topic constituted a meaningful conversation. Therefore, we set the lower threshold for calculating determinism at three messages (i.e., $lmin = 3$). The input data for computing entropy are the probability distributions generated by the determinism calculations.

Because we use repeated measures across performance episodes in our analyses, teams with data for only one episode have been removed from the dataset. The resulting sample sizes are $n = 126$ unrestricted teams and $n = 114$ time pressured teams.

Analyses

Recurrence plots provide the visual representation of the rhythm of team communication over time. For each team, two recurrence plots were constructed in Matlab, one plot for each performance episode. In Fig. 4.1, we show the recurrence plots for two teams as examples. The recurrence plots for all teams in the dataset can be found online in Mueller et al. (2019). Here we describe how to interpret the recurrence plots.

Note that both the *x*- and *y*-axes of these plots represent a Time Index, which results in the plots being symmetrical around a diagonal line moving out from the origin. The shaded areas indicate when systems revisit a given state at another point in time and density of those shaded areas facilitate quick assessments about the rhythm of the system. Specifically, congested graphics suggest that a system spends most of its time in a single state. Alternatively, a sparse graphic represents more shifts among states over time. For our investigation of the rhythm of communication, the visualizations depict periods of prolonged discussions (i.e., congested graphics) and short bursts about various topics (i.e., sparse graphics). Figure 4.1a depicts Distributed Team 1106 during the first performance episode. As can be seen

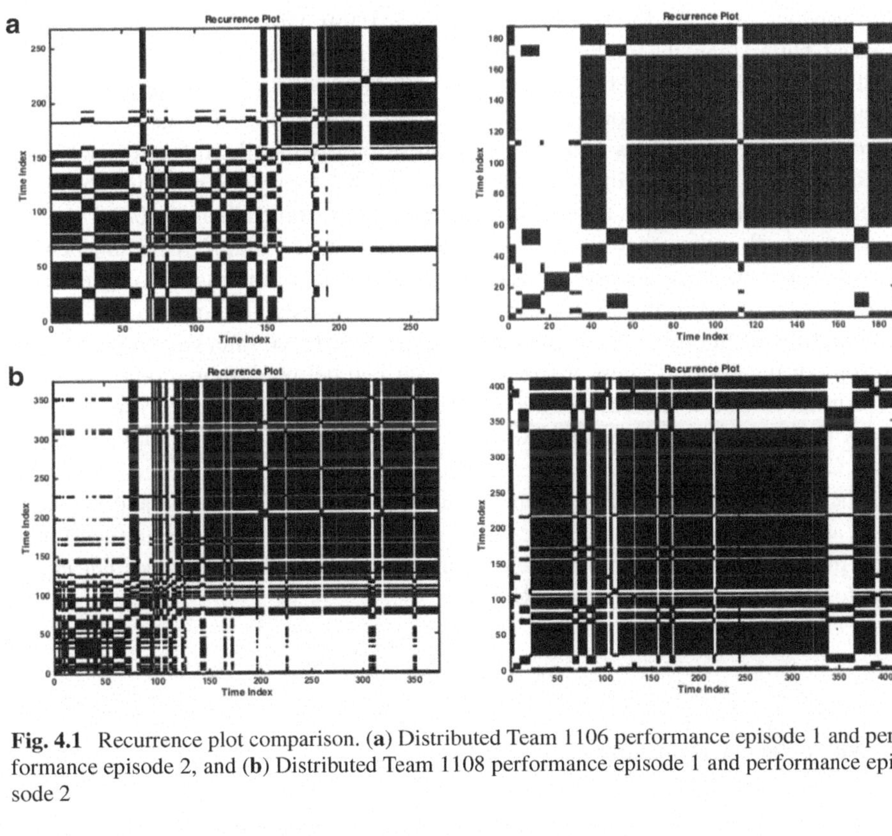

Fig. 4.1 Recurrence plot comparison. (**a**) Distributed Team 1106 performance episode 1 and performance episode 2, and (**b**) Distributed Team 1108 performance episode 1 and performance episode 2

in the figure, their conversation started with some switching among topics, which is depicted by the checkboard configuration in the lower left corner of the plot. They then transition to more prolonged discussions later, as depicted by the more densely shaded areas in the upper right corner. This rhythm is replicated in the second performance episode, but at a more accelerated pace as the team shifts to more focused conversation earlier.

Comparisons across teams also provide valuable insight. For example, the short bursts of topic switching exhibited by Distributed Team 1106 at the beginning of the first performance episode may not be that disjointed. This perspective comes from comparing the pattern to the very sporadic pattern representing Distributed Team 1108's early conversation during the first performance episode (see Fig. 4.1b). Distributed Team 1108's rhythm is a much more chaotic switching of topics depicted by the more fragmented checkerboard layout in the bottom left corner of the plot. Even during the later portion of the first performance episode, the plot is somewhat fragmented suggesting that this team may not have been able to adequately find their rhythm. They do appear to be more focused during the second performance episode.

Further comparisons of the recurrence plots may also be beneficial. For instance, we can compare the plots of teams with different levels of performance. When compared to the median performance score for both performance episodes (Median $\text{Performance}_{\text{T1}} = 0.83$; Median $\text{Performance}_{\text{T2}} = 0.99$), Distributed Team 1106 performed better than the median ($\text{Performance}_{\text{T1}} = 0.86$; $\text{Performance}_{\text{T2}} = 1$) and Distributed Team 1108 performed worse than the median ($\text{Performance}_{\text{T1}} = 0.60$; $\text{Performance}_{\text{T2}} = 0.92$). Considering that the differences in performance may be a function of the differences in team communication rhythm, the more focused conversations that transpired in Distributed Team 1106's communication may be more effective than the sporadic approach exhibited by Distributed Team 1108.

Not only do recurrence plots provide a means of visualizing recurrent behaviors, researchers have generated metrics calculated from the matrix of zeros and ones that facilitate comparisons across systems of interest (Eckmann et al. 1987). The metrics of interest for our research are the recurrence rate, determinism, and entropy of the system. These metrics are described below. A more comprehensive description of these metrics can be found in our introduction to recurrence analysis, with a specific emphasis on how it can be applied to group dynamics in Knight et al. (2016).

The recurrence rate represents the stability of a system over time and is calculated as the percentage of recurrence (i.e., repeat of the same code) in a recurrence plot. In other words, it quantifies the proportion of the plot that is shaded. The equation to calculate recurrence rates (RR) is:

$$RR = \frac{1}{N^2} \text{sum}_{i,j=1}^{N} R_{i,j}$$

Where the line of identity is excluded

Higher recurrence rates suggest prolonged periods in single states and correspond to congested graphics. Conversely, lower recurrence rates occur when graphics are sparse suggesting that the system changes states frequently.

Determinism captures the predictability of a system and can be seen graphically when diagonal line structures run parallel to the main diagonal line (Webber and Zbilut 1994). These diagonal line structures represent recurring topic cycles over time. The equation to calculate determinism (DET) is:

$$DET = \frac{\sum_{l=lmin}^{N} lP(l)}{\sum_{l=l}^{N} lP(l)}$$

Higher determinism represents a more predictable pattern. Alternatively, lower determinism is indicative of a system with a more random, dynamic process.

Entropy measures the complexity of the system's determinism by assessing the lengths of the upward sloping diagonal lines that represent determinism (Marwan et al. 2007; Webber and Zbilut 1994). It is quantified by calculating the Shannon entropy of the probability distribution of diagonal line lengths. The equation to calculate entropy (ENTR) is:

$$ENTR = -\sum_{l=lmin}^{N} p(l) \ln p(l)$$

The more uniform the structure, the lower the entropy. Higher entropy occurs when a system has a complex deterministic structure suggesting that conversation lengths vary over time.

Once all of the RQA metrics were calculated, we conducted linear mixed model with repeated measures regression analyses to assess the relationships among the constructs of interest. To facilitate these comparisons, we mean centered predictor variables. These analyses were conducted in SAS 9.4.

Findings for Team Communication Rhythm

The descriptive statistics and correlations are reported in Table 4.1. In Table 4.2 the results of our statistical analysis of recurrence quantification analysis (RQA) metrics are reported for both unrestricted and time pressured teams. Table 4.3 the results depicting the relationships between metrics and cost performance outcomes are provided; again results for both unrestricted and time pressured teams are included in the table.

Table 4.1 Rhythm descriptive statistics and correlations

Variables	Unrestricted *M*(*SD*)	Time pressured *M*(*SD*)	1	2	3	4	5	6	7	8
1. Medium	0.44 (0.50)	0.47 (0.50)	-	0.0001	0.05	0.68[a]	0.24[a]	0.48	0.38[a]	0.10
2. Episode	0.50 (0.50)	0.50 (0.50)	0.0001	-	0.94[a]	-0.19[a]	0.52[a]	0.29[a]	0.34[a]	0.63[a]
3. Minutes	100.70 (24.79)	125.03 (7.19)	0.36[a]	0.48[a]	-	-0.14	0.56[a]	0.32[a]	0.38[a]	0.63[a]
4. Messages	230.14 (138.53)	241.18 (169.56)	-0.04	-0.25[a]	-0.41[a]	-	0.26[a]	0.50[a]	0.44[a]	0.07
5. Recurrence rate	53.61 (15.26)	54.04 (15.48)	0.36[a]	0.28[a]	0.23[a]	0.18[a]	-	0.62[a]	0.65[a]	0.56[a]
6. Determinism	89.80 (9.03)	91.47 (8.76)	0.28[a]	0.24[a]	0.08	0.28[a]	0.71[a]	-	0.76[a]	0.36[a]
7. Entropy	2.74 (0.55)	2.96 (0.68)	0.16	0.23[a]	-0.03	0.45[a]	0.69[a]	0.70[a]	-	0.44[a]
8. Cost	0.89 (0.09)	0.80 (0.15)	0.12	0.86[a]	0.41[a]	-0.19[a]	0.30[a]	0.32[a]	0.23[a]	-

[a] $p < 0.05$

$N_{\text{unrestricted}}$ = 126 matched teams across time; 63 unique teams with data from both performance episodes. Correlations are shown in the bottom triangle of the matrix.

$N_{\text{time pressured}}$ = 114 matched teams across time; 57 unique teams with data from both performance episodes. Correlations are shown in the upper shaded triangle of the matrix.

Episode: first performance episode = 0; second performance episode = 1

Medium: distributed teams = 0, in person teams =1

Table 4.2 Analysis of recurrence metrics

	Unrestricted teams			Time pressured teams		
Variable	Model 1: Recurrence rate	Model 2: Determinism	Model 3: Entropy	Model 4: Recurrence rate	Model 5: Determinism	Model 6: Entropy
Intercept	63.55[a]	95.34[a]	3.12[a]	63.21[a]	96.21[a]	3.30[a]
Medium	-10.08[a]	-5.57[a]	-0.24[a]	2.05	-3.16[c]	-0.05
Episode	-8.98[a]	-5.91[a]	-0.46[a]	-18.93[a]	-6.40[a]	-0.60[a]
Minutes	-0.15	-0.04	0.004			
Messages	0.06[a]	0.02[b]	0.002[a]	0.07[a]	0.02[b]	0.002[a]
Minutes × episode	0.12	0.08	-0.003			
Messages × episode	-0.04[b]	0.001	0.0001	-0.05[a]	0.008	-0.0001
AIC	1005.3	890.9	211.4	884.5	770.3	219.2
Residual	114.09	67.01	0.20	116.75	45.20	0.28

[a] $p < 0.01$; [b] $p < 0.05$; [c] $p < 0.10$

$N_{\text{unrestricted}}$ = 126 matched teams across time; 63 unique teams with data from both performance episodes

$N_{\text{time pressured}}$ = 114 matched teams across time; 57 unique teams with data from both performance episodes

Episode: first performance episode = 0; second performance episode = 1

Medium: distributed teams = 0, in person teams = 1

Unstandardized coefficients reported.

AIC = Akaike Information Criterion

Table 4.3 Analysis of recurrence metrics and cost performance

Variable	Unrestricted teams cost performance Model 1	Model 2	Model 3	Time pressured teams cost performance Model 1	Model 2	Model 3
Intercept	0.99[a]	0.92[a]	0.98[a]	0.89[a]	0.79[a]	0.33
Medium	-0.03[b]	-0.02[b]	-0.02[c]	0.03	0.12	0.03
Episode	-0.17[a]	-0.17[a]	-0.25[b]	-0.20[a]	-0.15[a]	0.56[b]
Minutes	0.0001	0.0001	-0.0001			
Messages	-0.0001	-0.0001	-0.0001	0.001[a]	0.0001	0.0001
Recurrence rate		-0.0001	0.0001		0.002[b]	0.002
Determinism		0.001[c]	-0.0001		-0.002	0.004
Entropy		-0.01	0.003		0.03	0.003
Minutes × episode			0.0001			
Messages × episode			0.0001			0.0001
RR × episode			-0.001			-0.001
DET × episode			0.003[b]			-0.01[b]
ENTR × episode			-0.06[b]			0.08
AIC	-360.2	-329.4	-276.0	-127.8	-121.3	-89.4
Residual	0.002	0.002	0.002	0.009	0.009	0.009

[a]$p < 0.01$; [b]$p < 0.05$; [c]$p < 0.10$

$N_{unrestricted}$ = 126 matched teams across time; 63 unique teams with data from both performance episodes

$N_{time\ pressured}$ = 114 matched teams across time; 57 unique teams with data from both performance episodes

Episode: first performance episode = 0; second performance episode = 1

Medium: distributed teams = 0, in person teams = 1

Unstandardized coefficients reported.

AIC = Akaike Information Criterion

Visualizing Conversation Rhythm

With respect to RQ 4.1, we examine the recurrence plots to identify rhythmic patterns in the communication and ascertain any differences in the patterns exhibited by high and low performing teams. Two sample plots, one for a high and one for a low performing team, are shown in Fig. 4.1. The recurrence plots for all teams can be examined in Mueller et al. (2019). First, the recurrence plots suggest that teams generally have a rhythm of focusing on topics and moving on to other topics, thus creating boxes across the plots. As teams repeat this process on other topics, a checkerboard pattern is created. The density or sparseness of the plot depicts the amount of cycling on topics and changing among topics that happens. As well, the size of the boxes in the checkboard provides a visual indication of the rhythm teams had in discussing topics over time.

Second, the comparison of teams achieving good cost performance, such as Team 1106, versus those achieving lower cost performance, such as Team 1108, suggests an interesting difference; teams with better cost performance generally have a more defined rhythm. That is, the communication patterns have clearer boxes and a more sparse checkerboard pattern suggesting that teams focused discussions on a specific topic before moving on to another topic. Moreover, differences in early and late patterns that become more pronounced are evident, especially in teams that achieve good cost performance over time. That is, the rhythm of progressing through topics shows the ebb and flow of topic discussions, and for teams with better cost performance, larger boxes generally appear early in conversation suggesting more focused discussions around one or a couple of topics, before shifting to other topics. This visual evidence provides a preliminary response to RQ 4.4.

Examining Conversation Rhythm Using Recurrence Metrics

Rhythm Differences Across Time Constraints and Communication Media

Rhythm Differences Across Time Constraints and Communication Media
Results

Our second and third research questions focus on differences in communication rhythm among teams working with different constraints (i.e., unrestricted and time pressured) and communicating via different media (i.e., distributed and in person), respectively. We use the results of our statistical analysis of recurrence metrics (see Table 4.2) to answer these questions. Specifically, we assessed the way time constraints and communication media may have influenced the recurrence rate, determinism, and entropy that teams attained. For unrestricted teams, the recurrence rate (see Model 1) was found to be lower for teams meeting in person ($\beta = -10.08$, $p < 0.01$) and lower during the second performance episode ($\beta = -8.98$, $p < 0.01$). The recurrence rate increased slightly by the number of messages conveyed by the team ($\beta = 0.06$, $p < 0.01$) yet the effect of messages was moderated by episode ($\beta = -0.04$, $p = 0.03$) and therefore the higher the number of messages the higher the recurrence rate, and the increase was greater in the first performance episode. The findings for determinism (Model 2) and entropy (Model 3) are similar; in both cases, the metrics are lower when teams meet in person (determinism medium $\beta = -5.57$, $p < 0.01$; entropy medium $\beta = -0.24$, $p < 0.01$) and lower in the second performance episode (determinism episode $\beta = -5.91$, $p < 0.01$; entropy episode $\beta = -0.46$, $p < 0.01$). Also, determinism and entropy increase as the number of messages increase (determinism messages $\beta = 0.02$, $p = 0.01$; entropy messages $\beta = 0.002$, $p < 0.01$).

For the assessment of time pressured teams' RQA metrics, our patterns of results are similar to the aforementioned patterns associated with unrestricted teams with one exception. In Model 4 of Table 4.2, the recurrence rate was found to be lower during the second performance episode ($\beta = -18.93$, $p < 0.01$). As well, the recurrence rate increased by the number of messages conveyed by the team ($\beta = 0.07$, $p < 0.01$) and the effect of messages was moderated by performance episode ($\beta = -0.05$, $p < 0.01$) indicating that the higher the number of messages the higher the recurrence rate, and the increase was greater in the first performance episode. Interestingly, the medium of interaction (i.e., distributed versus in person) was not significant for time pressured teams ($\beta = 2.05$, $p = 0.51$), whereas it was significantly different for unrestricted teams. The determinism (Model 5) and entropy (Model 6) for teams working under time pressure unfolded in a similar manner to unrestricted teams. That is, determinism was lower in the second performance episode ($\beta = -6.40$, $p < 0.01$) and increased with the number of messages exchanged ($\beta = 0.02$, $p = 0.02$). The effects for entropy were similar (episode $\beta = -0.60$, $p < 0.01$; messages $\beta = 0.002$, $p < 0.01$).

Overall Patterns Observed in Rhythm Differences Across Time Constraints and Communication Media

We now extrapolate the patterns that emerge from our results. The results confirm that the rhythm of communication was very similar for unrestricted and time pressured teams, with the exception of the difference observed between the recurrence rates of distributed and in person teams with unrestricted time to complete their task. Interesting patterns associated with the similarities and differences in these results are worth noting, and may be useful in bolstering our understanding of communication rhythm.

Our results suggest that when working in an unrestricted setting, in person teams may be changing topics more often and therefore may have a faster rhythm than distributed teams. Time pressured teams, however, exhibited similar recurrence rates regardless of the media through with they were communicating. In other words, the added pressure of having limited time to complete their task may outweigh the differences in communication media when teams consider how long they can discuss various topics.

The remaining noteworthy patterns of results were similar for both unrestricted and time pressured teams. Specifically, the rhythm of communication is dependent upon how much communication occurs (i.e., the number of messages) and how many times the team has worked together (i.e., the performance episodes). With respect to the changes from one performance episode to the next, these results suggest that teams may have engaged in more focused conversations about topics during the first performance episode than in the second. But, at the same time, they may have covered more territory as they quickly iterated among the topics. That is, the increased recurrence rates suggest that teams were engaged in prolonged discussions, the determinism metrics indicate that they were revisiting topics in focused ways, and entropy reveals that they also were changing topics. In other words, teams may have been rapidly cycling through a series of in depth discussions about different topics during the first performance episode. Conversely, in the second episode they may have moved more quickly and purposefully through the topics since they were already familiar with the basics of the task they were assigned.

As the number of messages exchanged increased, the recurrence rate, determinism, and entropy all increased. An increased volume of communication, therefore, appears to increase a team's ability to conduct more in depth conversations about many topics in more predictable cycles. Interestingly, the positive relationship between the numbers of messages exchanged and the recurrence rate is more pronounced during the first performance episode than the second. This finding suggests that teams may start the first performance episode with rhythmic cycles focused on transition processes (e.g., mission analysis, tactical strategy) as they work to comprehend what they have been tasked with doing. As they transition into the second performance episode, they have already successfully completed the task once and, therefore, can focus more of their conversation on doing the task with periodic bursts of communication about the transition processes to ensure they are still on track.

Rhythm Differences Across Performance

Rhythm Differences Across Performance Results

For our fourth research question we investigated the relationship between recurrence metrics and cost performance. In this analysis, cost performance is scaled to be better when higher (see Chap. 2 for a description of the transformation used to create this variable). The results for unrestricted teams are reported in Model 3 in Table 4.3. Cost performance was lower in the second performance episode ($\beta = -0.25$, $p = 0.03$). Moreover, two significant moderated relationships were found and are plotted in Fig. 4.2 for easier interpretation. First, the relationship between determinism and cost performance differs by episode ($\beta = 0.003$, $p = 0.05$) and is depicted graphically in Fig. 4.2a. In the first performance episode, the relationship between determinism and cost performance is constant, whereas in the second performance episode, as determinism increases, performance increases. Second, the relationship between entropy and cost performance is moderated by episode ($\beta = -0.06$, $p = 0.04$) and is shown in Fig. 4.2b. As with determinism, the relationship between entropy and cost performance was constant during the first performance episode. The relationship between these two variables during the second performance episode suggests that as entropy increases performance decreases.

For teams working under time pressure (see Model 3 in Table 4.3), we find that cost performance was higher in the second performance episode ($\beta = 0.56$, $p = 0.05$). Further, we find that the relationship between determinism and cost performance is moderated by episode ($\beta = -0.01$, $p = 0.03$). This significant moderated relationship (shown in Fig. 4.2c) suggests that as determinism increases, performance increases during the first performance episode. Conversely, as determinism increases in the second performance episode, cost performance decreases.

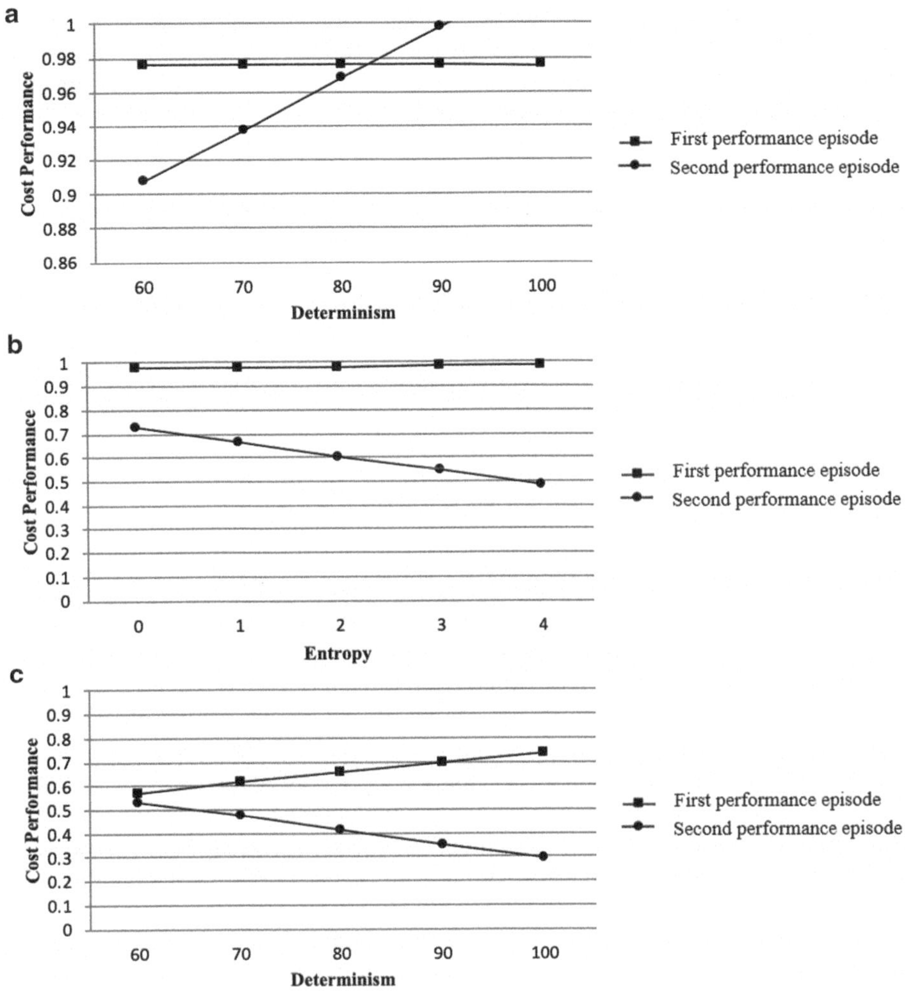

Fig. 4.2 Recurrence analysis significant moderated relationships. (**a**) Unrestricted teams, determinism and cost performance, (**b**) Unrestricted teams, entropy and cost performance, and (**c**) Time pressure teams, determinism and cost performance

Overall Patterns Observed in Rhythm Differences Across Performance

Our results suggest, in answer to our fourth research question, that high and low performing teams do exhibit some differences in their communication rhythms. Interestingly, cost performance was higher in unrestricted teams' first performance episodes and higher in time pressured teams' second performance episodes.

When focusing on the relationships between recurrence metrics and cost performance we found that determinism and entropy mattered, although the effects were different depending on the time constraints they faced. Neither determinism nor entropy appear to make a difference in unrestricted teams' abilities to perform their tasks during the first performance episode. During the second performance episode, however, they may benefit from engaging in more uniform, predictable patterns of communication, as these characteristics were related to better cost performance. Time pressured teams may also benefit from more predictable patterns of communication, but during the first performance episode. During the second performance episode these teams may find more random, dynamic conversations advantageous. These dramatic differences between time availability and across performance episodes detail how malleable teams must be in order to adapt to the changing demands of the conditions under which they are collaborating.

Theoretical Implications and Future Research

The two theoretical frameworks unpinning our research, Marks et al.'s (2001) recurring phase model of team processes and Bartunek and Woodman's (2015) temporal dimensions of change, recognize the need for more sophisticated representations of the complexities faced by teams. Marks and colleagues advocate for expanding upon the traditional input-process-output model by incorporating the realistic cycles that occur both sequentially and simultaneously. Similarly, Bartuek and Woodman contend that change does not occur linearly as suggested by Lewin's (1947) unfreeze-change-refreeze model, which has dominated the field of organizational development for half a century. Our findings from recurrence quantification analysis (RQA) underscore the need for this shift toward capturing the complex realities of team behavior.

Marks and colleagues (2001: 359) also suggest that team "rhythm" is "marked by identifiable periods of action and transition periods between actions" that can be identified in our recurrence plots. Moreover, they predict repetitive occurrences of different processes over time as teams transit through cycles of focused and chaotic conversation. Bartunek and Woodman (2015) predict that these repetitive cycles will vary in intensity and frequency over time. Support for these theoretical notions is evident in our results. Indeed, the most prominent differences observed exist between the two performance episodes, regardless of the contextual conditions in which the teams are working. For example, all teams appear to rapidly cycle through focused conversations about the topics during the first performance episode and then shift to more sporadic, short bursts of communication during the second performance episode. The nature of these short bursts may differ, however, between unrestricted and time pressured teams, as demonstrated by our better performing teams in each condition. Specifically, unrestricted teams' bursts occurred more predictably and time pressured teams engaged in more random, dynamic bursts.

Future researchers may benefit team science by continuing to examine teams using methods such as RQA as the results reflect the realities of team dynamics. These types of inquiries may lead to the identification of indicators that could be used to monitor communication rhythms in real time, and offer corresponding guidance for facilitating team communication in a way that balances forward progress with the need to cyclically iterate among transition and action processes. One such burgeoning area of inquiry employs neurological and physiological measures. By using electroencephalogram data, researchers have second-by-second measurements of brain rhythms. Researchers have used this approach to examine the neurophysiologic synchronies and co-expressions of activities during dyad and group work (Delorme et al. 2002; Stevens and Galloway 2016). Other researchers are employing physiological measures to identify team-level phenomena. For example, Mønster et al. (2016) examined physiological synchrony among team members and identified cohesion through synchrony of smiling and conflict/tension through synchrony of skin conductance. Triangulation among constructs such as team behaviors, communication, cognition, neurodynamics, and physiological measures is needed. RQA may be one of the many tools that can be employed to distinguish how neurological and physiological data correspond to team behaviors. The advances in gathering neurological and physiological data unobtrusively suggest that the time is right for such action within the research community.

Methodological Contributions and Future Research

From a methodological perspective, our results in this chapter demonstrate how useful visualizations can be in capturing the continuous flow of activity over time, which in turn informs our understanding of team progression. These visualizations, however, typically represent individual team dynamics. To garner a more holistic picture of these dynamics that incorporates a larger sample, we employ three metrics (i.e., the recurrence rate, determinism, and entropy metrics), which are relatively new to the field of team behavior. These metrics facilitate drawing inferences about how communication transpires over time across the entire sample.

Recurrence quantification analysis (RQA) has its roots in physics and has been applied to a variety of continuous processes such as monitoring heart rhythms. Team researchers may benefit from identifying other methodological tools used in disciplines outside the social sciences where continuous data is routinely analyzed. Not only will this approach result in bringing new tools into team science, it may also provide opportunities to engage in interesting interdisciplinary collaborations. As Salas et al. (2018) point out, such interdisciplinary collaborations have great potential to advance team science by addressing more real-world problems and help close the gap between what researchers learn through their investigations and what practitioners use in their day-to-day activities.

Managerial Minutes

Time pressure calls for agile communication management.
Time pressured teams may feel rushed as they work to understand what they have been asked to do and determine how they are going to do it. Under these constraints, you may want to consider an iterative approach, where periodic bursts of conversation focus on whether or not teams still understand their tasks, believe their approaches are best, and are on track to successfully complete their assignment.

Iteration among topics is necessary, but varies over time; be flexible.
In person teams may bounce around topics during the early stages of their time together. You may want to help these teams shift toward more uniform, predictable conversations as they become familiar with what they are doing as a team.

Appendix: How to Conduct Recurrence Analysis

In Chap. 4, we examined the rhythm of communication by visually assessing the recurrence plots. Additionally, we calculated the recurrence quantification analysis (RQA) metrics and statistically analyzed them using repeated measure linear regression analysis. In this appendix, we demonstrate how the results and findings are generated using the example dataset introduced in the Chap. 3 Appendix. A more comprehensive treatment of recurrence analysis can be found in Knight et al. (2016).

The recurrence plots and metrics were generated using Matlab, including the recurrence rate, determinism, and entropy. The statistical analyses were conducted in SAS 9.4.

Data Preparation

Matrix Generation

To conduct the analyses, the message topic codes and performance data are used. To generate the recurrence plots for each team, a square matrix is generated that corresponds to the length of the communication string. For example, the first team in our example dataset has a communication string with 20 exchanges. The topic codes are:

10, 10, 10, 20, 20, 20, 20, 30, 30, 30, 10, 10, 10, 40, 40, 40, 40, 40, 40, 40

Table 4.4 Recurrence matrix for example dataset Team 1

40	0	0	0	0	0	0	0	0	0	0	0	0	0	1	1	1	1	1	1	1
40	0	0	0	0	0	0	0	0	0	0	0	0	0	1	1	1	1	1	1	1
40	0	0	0	0	0	0	0	0	0	0	0	0	0	1	1	1	1	1	1	1
40	0	0	0	0	0	0	0	0	0	0	0	0	0	1	1	1	1	1	1	1
40	0	0	0	0	0	0	0	0	0	0	0	0	0	1	1	1	1	1	1	1
40	0	0	0	0	0	0	0	0	0	0	0	0	0	1	1	1	1	1	1	1
40	0	0	0	0	0	0	0	0	0	0	0	0	0	1	1	1	1	1	1	1
10	1	1	1	0	0	0	0	0	0	0	1	1	1	0	0	0	0	0	0	0
10	1	1	1	0	0	0	0	0	0	0	1	1	1	0	0	0	0	0	0	0
10	1	1	1	0	0	0	0	0	0	0	1	1	1	0	0	0	0	0	0	0
30	0	0	0	0	0	0	0	1	1	1	0	0	0	0	0	0	0	0	0	0
30	0	0	0	0	0	0	0	1	1	1	0	0	0	0	0	0	0	0	0	0
30	0	0	0	0	0	0	0	1	1	1	0	0	0	0	0	0	0	0	0	0
20	0	0	0	1	1	1	1	0	0	0	0	0	0	0	0	0	0	0	0	0
20	0	0	0	1	1	1	1	0	0	0	0	0	0	0	0	0	0	0	0	0
20	0	0	0	1	1	1	1	0	0	0	0	0	0	0	0	0	0	0	0	0
20	0	0	0	1	1	1	1	0	0	0	0	0	0	0	0	0	0	0	0	0
10	1	1	1	0	0	0	0	0	0	0	1	1	1	0	0	0	0	0	0	0
10	1	1	1	0	0	0	0	0	0	0	1	1	1	0	0	0	0	0	0	0
10	1	1	1	0	0	0	0	0	0	0	1	1	1	0	0	0	0	0	0	0
	10	10	10	20	20	20	20	30	30	30	10	10	10	40	40	40	40	40	40	40

Therefore, a 20×20 matrix is constructed. The matrix for Team 1 is given in Table 4.4.

To construct the matrix, start by labeling the rows and columns with the topic codes in sequential order starting in the lower left corner. For example, the first row and column represent the first exchanged message, which is coded as 10. The matrix is completed by assigning 1 s to indicate when the topic codes match and 0 s otherwise. For example, since the topic code 10 is the code associated with the first row, every column that represents a message coded as topic code 10 is designated as a 1, all other topic codes are designated 0. Once complete, the matrix is used for visualization and for calculating the recurrence quantification metrics.

Recurrence Quantification Analysis Metrics Calculations

Three RQA metrics are used in our analyses, namely recurrence rates, determinism, and entropy. These metrics for all teams in the example dataset are given in Table 4.5.

Table 4.5 Recurrence quantification analysis metrics for the example dataset teams

Example team	Performance	Recurrence rate	Determinism	Entropy
1	0.98	23.68	53.33	1.24
2	0.85	22.08	43.14	0.41
3	0.75	24.28	44.78	1.04
4	0.90	21.65	34.00	0.67
5	0.40	22.83	20.63	0.56
6	0.65	23.38	37.04	1.05
7	0.30	31.05	55.93	0.97
8	0.40	23.38	35.19	0.95
9	0.22	25.97	23.33	0.56
10	0.30	25.97	20	0.64

Recurrence Rate

The recurrence rate (RR) measures the percentage of times a team revisits a topic. To calculate the recurrence rate, we sum the number of identified matches (i.e., the 1 s in the matrix) minus the number of points on the main diagonal (i.e., the line of identity), and multiply the sum by 1 divided by the string length squared minus the number of points on the main diagonal. The equation for recurrence rate is:

$$RR = \frac{1}{N^2}\text{sum}_{i,j=1}^{N} R_{i,j}$$

Where the line of identity is excluded

The recurrence metrics for all teams in the example dataset are provided in Table 4.5. Looking specifically at Team 1, the total number of 1 s across the matrix is 110. We then subtract the number of points on the main diagonal which is 20 to equal 90. We then take the string length (n = 20) squared which is 400 and subtract the number of points on the main diagonal which is 20 to equal 380. As such, the recurrence rate for the example team is 90/380 = 23.68%. The interpretation of this number indicates that the team revisited the same topic for about 23% of the conversation.

Determinism

Determinism (DET) is an indicator of a system's predictability. This metric represents the percentage of recurrence (i.e., repeat of same code) along upward-running diagonal lines in a recurrence plot that are parallel to the main diagonal running from the bottom left to top right corner. These upward-running diagonal lines represent periods of sustained communication about a particular topic.

The calculation calls for looking at the frequency distribution of the diagonal line lengths, where the diagonal line length represents continued repetition in a state. The researcher needs to determine what constitutes enough repetition to be meaningful and set a minimum repetition number (for more information about minimum repetition numbers see Marwan and colleagues' 2007). In our case, as described in Chaps. 3 and 4, we inductively ascertained that at least three consecutive messages about a single topic constituted a meaningful conversation. Therefore, we set the number to 3 (i.e., *lmin* = 3), as it represents a focused discussion on a topic. Determinism is calculated as the distribution of recurrent points along diagonal lines that have met our minimum repetition threshold. The equation for determinism is:

$$DET = \frac{\sum_{l=lmin}^{N} lP(l)}{\sum_{l=l}^{N} lP(l)}$$

In our analysis of example Team 1, the determinism is 52.50%. The interpretation of this number indicates that teams engage in meaningful conversations more than half of the time, where meaningful conversations are defined as three or more consecutive messages about the same topic. Moreover, as shown in Table 4.5, Team 1 is more predictable than the other teams because it has the highest determinism score.

Entropy

Entropy (ENTR) is as assessment of a system's complexity. The calculation reflects Shannon entropy, and is based upon the probability distribution of the upward-running diagonal line lengths. In our dataset, it helps us understand how frequently teams are moving among states or topics across the conversation.

The entropy value is not a percentage like the other metrics, but rather can be used as a comparative scale. The calculation determines the likelihood of the upward-running diagonal lines being of given lengths in the recurrence plot. The equation to calculate entropy is:

$$ENTR = -\sum_{l=lmin}^{N} p(l)\ln p(l)$$

Team 1 from the sample dataset has an entropy value of 1.33, which can be used to compare the complexity of the conversations in which Team 1 engaged to those of other teams. For example, when compared to Team 2 (ENTR = 0.41), we can see that Team 1 engaged in more complex conversations. In other words, Team 1's conversations were more dynamic, whereas Team 2's conversations were more uniform.

Data Usage

Recurrence Plots

The matrix created with 1 s and 0 s can be converted into a plot that visually represents the recurrence of states across a system. Simply by turning 1 s into a darkened cell and 0 s into a blank cell, the recurrence plot is rendered. These recurrence plots help to visualize differences in recurrence rates, determinism, and entropy.

Sample recurrence plots are shown in Fig. 4.3, with Team 1 in Fig. 4.3a, Team 2 in Fig. 4.3b, and Team 3 in Fig. 4.3c. Through visual inspection we may get a sense of the differences across teams; specifically, Team 2 has a lower recurrence rate than Team 1, which in turn is lower than Team 3. While for determinism, Team 2 and Team 3 are lower than Team 1. Interpretatively, picking up on key differences, one might say that Team 1 behaves more predictability throughout the conversation, notably shown through having more diagonal line structures in the recurrence plots. Team 2 was less repetitive, which yields a sparser picture, whereas Team 3 may be considered denser. Finally, Team 1 has the highest complexity, which suggests the team was moving among topics more than other teams as can be seen in the more sporadic plot.

Regression Analysis

To assess the impact of recurrence metrics on an outcome variable we use SAS 9.4 to conduct linear regression analyses. In Chap. 4, we used a repeated measures approach for our analyses because our data had two performance episodes. Our example dataset, however, is much simpler for illustration purposes. Therefore, we apply linear regression to assess how the independent variables of recurrence rate, determinism, and entropy relate to the dependent variable team performance. The results of our analysis are shown in Table 4.6. These results indicate that two recurrence metrics are significantly related to performance. First, recurrence rate is negatively related to performance ($\beta = -0.084$, $p < 0.01$). This relationship suggests that the more frequently teams revisit topics, the worse they will perform. Second, determinism is significantly and positively related to performance ($\beta = 0.016$, $p = 0.01$). In other words, teams with more predictable communication patterns may have better performance.

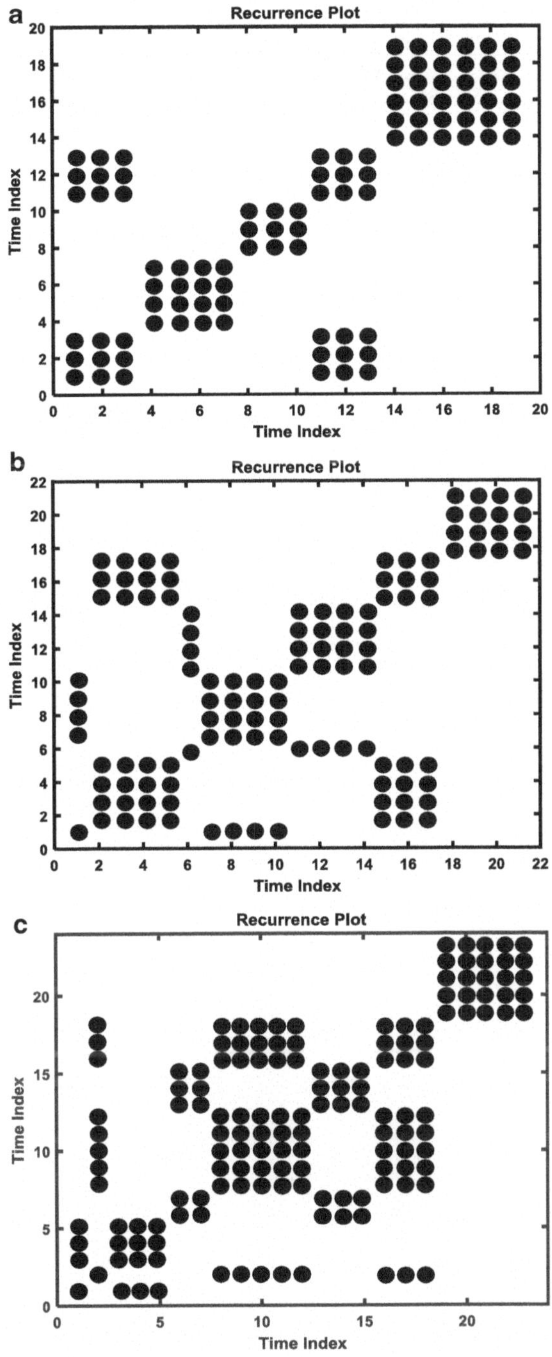

Fig. 4.3 Recurrence plots for example dataset. (**a**) Example Team 1 recurrence plot, (**b**) Example Team 2 recurrence plot, and (**c**) Example Team 3 recurrence plot

Table 4.6 Correlation and regression analysis results for example dataset recurrence metrics and performance

Correlation results					Regression results	
Variable	1	2	3		Variable	Performance model
1. Performance	--				Intercept	2.058[a]
2. Recurrence rate	-0.62[c]	--			Recurrence rate	-0.084[a]
3. Determinism	0.50	0.27	--		Determinism	0.016[b]
4. Entropy	0.27	0.21	0.63[b]		Entropy	-0.031
					F-statistic	12.251[a]

[a] $p < 0.01$; [b] $p < 0.05$; [c] $p < 0.10$

Chapter 5
Timing and Pacing of Team Communication

Whether a team is tasked with a developing a new product, incrementally improving an existing situation, or reacting to an emergency, all members of the project team must collectively discuss the what they need to be doing, how they will do it, and when it will happen. Meetings to discuss these topics will occur periodically throughout the teams' lifecycles. Team leaders can help facilitate this process by planning what, and when, various topics will be discussed. The results from our focus on communication timing and pacing in this chapter will help inform these plans. Timing identifies what topics a team may benefit from discussing early, such as the parameters or constraints of the project, versus those that may be more appropriately considered later, such as specific work assignments. Pacing will help team leaders consider when and how much teams may need to iterate among topics during these meetings at various times over their lifecycles. Early conversations about what the team is tasked with accomplishing and creating a strategy for accomplishing it may require longer and/or more frequent meetings. Over time, team meetings may become shorter or less frequent as the members can better use their time accomplishing the tasks assigned them. The specific topics and timing may differ depending upon the conditions under which the team is working. While these seem like simple considerations, rarely are they employed when setting meeting schedules and agendas. Understanding the timing and pacing of conversations will be useful in creating new heuristics for team leaders as they plan for effective project team communication.

Introduction to Team Communication Timing and Pacing

In Chap. 4, we ascertained that team communication is comprised of rhythmic patterns. For our initial foray into understanding what is happening within those patterns, we employ visualization and simple statistics to examine the timing and

S. McComb, D. Kennedy, *Computational Methods to Examine Team Communication*, Computational Social Sciences,
https://doi.org/10.1007/978-3-030-36159-4_5

pacing of the conversations transpiring among team members. This approach provides a basic understanding of how team communication is unfolding over time. In their conceptualization of organizational change, Bartunek and Woodman (2015) suggest that timing and pacing depict when activities occur and the speed of change, respectively. We extrapolate these definitions to the team domain by examining when specific topics are discussed to capture timing and how often teams change topics to assess pacing. Moreover, we compare what is occurring across quartiles to determine if timing and pacing change over time.

Starting with this basic approach has proven useful in past group and team research. For example, Olson and colleagues (1992) examined the proportion of time groups spent on specific activities during software design meetings and how they shifted among those activities. Their findings identified iterative cycles between specific design activities and coordination activities. This finding provides evidentiary support for the conceptualization of team collaboration proffered by Marks and colleagues (2001) that is a foundation of this research. Specifically, teams iterate between action and transition processes, which include coordinating activities. Important for our purposes in this chapter is Olson et al.'s (1992) approach of simply providing a graphical representation of the time spent on various activities and the transitions among activities that transpired during the meetings observed. Their approach also aligns with that of other researchers who have examined the frequency and complexity of communication (e.g., Stachowski et al. 2009).

To examine communication timing and pacing, we consider the frequency with which the team speaks about specific topics (timing) and the complexity of transitions among those topics (pacing). Although Olson and colleagues (1992) examined both frequency and transitions among actions occurring at a single meeting, our efforts expand upon their research in two specific ways. First, we break the team's communication into quartiles. Several researchers have identified changes that occur in communication over time (e.g., Gersick 1988, 1989; Hewes and Poole 2012; Kennedy and McComb 2014; Pavitt 1992; Pavitt and Johnson 2001; Wallace 1987). For example, Gersick (1988) found that teams cycle through different phases with a specific shift occurring at the midpoint between their first meeting and the deadline. Our decision to examine communication within quartiles provides the opportunity to examine changes associated with the team's time together, but with a finer degree of granularity that may be necessary for capturing nuances (Ballard et al. 2008).

Second, we examine transitions using social network analysis. Applying network analysis techniques to examine the relationships among topics is not new to team research. For example, in shared mental model research, network measures are used to calculate the structural similarity among team members' mental models (e.g., Mathieu et al. 2000; 2005) and ascertain the centrality of particular concepts embedded in team members' mental models (e.g., Mathieu et al. 2000; 2005; Uitdewillinger et al. 2013). Our interest is in examining the speed of change, or pacing, that occurs during team conversations, which we operationalize as how often teams are transitioning among topics. Therefore, our interests are analogous to social network researchers' interests in how much interaction is occurring among network actors, which is characterized as network density. Density is calculated by averaging tie strength between nodes in the network diagram (Borgatti et al. 2018), where nodes are the network actors. In our data, nodes are the topics, tie strength is the number of transitions between nodes, and the density scores represent the complexity of transitions among topics.

In our assessment of team communication used to complete a workforce scheduling task, the topics of interest include action and different types of transition processes (i.e., goal specification, mission analysis, operational strategy, and tactical strategy). Researchers suggest that certain transition processes may be particularly salient earlier in taskwork, such as discussion of the task parameters (i.e., mission analysis) (Hackman et al. 1976; Woolley 1998), and setting a plan (i.e., tactical or operational strategy) (e.g., Tschan 1995; von Cranach et al. 1982; Weingart 1992). We believe these patterns can be realized through the assessment of timing and pacing. Yet, we further acknowledge that timing and pacing may be subject to the characteristics of the team environment. That is, the media used to carry the conversation may affect the flow of communication. Researchers suggests that lean media may limit information sharing (Daft and Lengel 1986) such that teams distributed across a networked environment may need to make more efficient use of communication. As well, working under time pressure disrupts normal processes (Driskell et al. 1999; Ellis 2006; Kelly and McGrath 1985) and teams may accelerate the timing and pacing of conversations out of necessity. Thus, we expect that differences exist in the timing and pacing of topics, and these effects will be impacted by the team's communication media (i.e., distributed versus in person) and time requirements (i.e., unrestricted time or time pressured).

Research Questions

The research questions associated with timing are:

RQ5.1. Does the frequency with which teams discuss different topics change over time?

RQ5.2. Does the frequency with which teams discuss different topics vary depending on the time they have available to complete their task (i.e., unrestricted time versus time pressured)?

RQ5.3. Does the frequency with which teams discuss different topics vary depending on the communication media available to them (i.e., distributed versus in person)?

The research questions associated with pacing are:

RQ5.4. Does the complexity of team conversations change over time?

RQ5.5. Does the complexity of team conversations vary depending on the time they have available to complete their task?

RQ5.6. Does the complexity of team conversations vary depending on the communication media available to them?

Methodology

Methodological Approaches: Network Pictographs and Statistical Comparisons

Network pictographs and statistical comparison tests were selected as the methodological approaches to assess timing and pacing. Network pictographs depict team communication activity that can be compared visually. Simple visualizations convert complex information into tractable illustrations that are easier to comprehend than tables of raw or summarized data, from which the observer may identify patterns and anomalies more readily.

Statistical comparisons output information about similarities and differences over time and across study conditions. The indication of significant differences, or the lack thereof, offers evidence supporting the patterns observed in the network pictographs, and, therefore, facilitates answering our research questions. Moreover, taken together these two methodological approaches provide basic information about longitudinal team communication that represents a logical next step in our holistic investigation reported in this book.

Data Specification

The communication strings must be manipulated in two ways to assess timing and pacing. First, to answer RQ5.1 through RQ5.3, the frequency counts of each topic during each quartile for each performance episode must be calculated. These counts are used to conduct the statistical tests necessary to examine timing. Second, to answer RQ5.4 through RQ5.6, the frequency counts of each $t + 1$ transition during each quartile for each performance episode is used to construct the network pictographs and calculate network density. Network density is used to examine pacing.

The sample sizes by type of team for both performance episodes are:

- $n_{T1} = 35$ and $n_{T2} = 35$ unrestricted distributed teams
- $n_{T1} = 35$ and $n_{T2} = 28$ unrestricted in person teams
- $n_{T1} = 30$ and $n_{T2} = 30$ time pressured distributed teams
- $n_{T1} = 28$ and $n_{T2} = 27$ time pressured in person teams.

Analyses

The visualizations were used to examine both timing and pacing. Specifically, network pictographs depicting the various topics available to discuss and the transitions among them were created for each condition (i.e., media and time availability)

Fig. 5.1 Example network pictograph

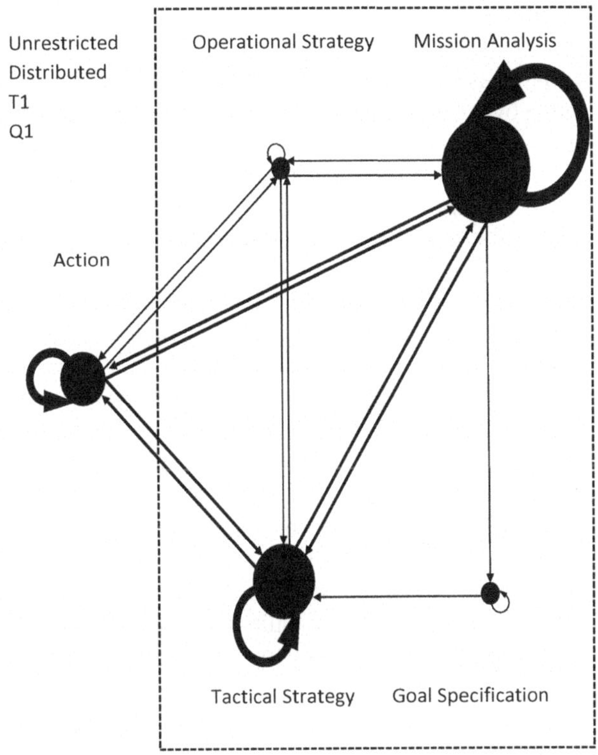

across the four quartiles. All network pictographs can be seen online in Mueller and colleagues (2019).

These diagrams were constructed in PowerPoint, frequency counts were scaled to balance creating an accurate representation of the data with the limitations of PowerPoint. An example network diagram for distributed teams' mean communication in the first quartile of the first performance episode is shown in Fig. 5.1. The transition processes are inside the box and the action process is outside. The circles depict communication topics, where the sizes of the circles represent the relative frequencies of messages exchanged about each topic. For example, the circle for mission analysis is larger than the circle for goal specification indicating more messages about mission analysis were transmitted. The arrows depict communication pacing, where the width of the arrow represents the relative frequencies of transitions from one topic to another. For instance, the transitions from mission analysis to mission analysis (i.e., the curved line outside the mission analysis circle) occurred more frequently than all other transitions. Where no arrows exist, no transitions occurred.

The analytic approaches used to answer the research questions associated with timing and pacing are described below. All analyses were conducted in SAS 9.4.

Timing

Timing was examined by comparing the frequency with which each topic was discussed over time and across conditions and media. To compare the frequencies over time, a repeated measures analysis of variance (ANOVA) was performed for each topic. In other words, we tested the hypothesis that the mean frequency with which a topic was discussed did not significantly differ across quartiles. Univariate tests of hypotheses for within subject effects were performed to ascertain if differences over time existed. Finally, profile contrasts were run to determine if the frequency in one quartile was significantly different than the frequency in the next quartile.

To ascertain any differences in frequency attributable to the team's time constraints or their communication media, a 2 × 2 full factorial ANOVA was run for each topic in each quartile for each performance episode (e.g., mission analysis in Quartile 1 during the first performance episode, operational strategy in Quartile 3 during the second performance episode). In other words, we tested the hypothesis that the mean frequency with which a topic was discussed did not significantly differ across time constraints or communication media, respectively. As the sample size of each cohort differed, the least squares mean procedure with a Tukey-Kramer test was run to ascertain significant differences in frequency across the four types of teams (i.e., distributed with unlimited time, distributed with time pressure, in person with unlimited time, and in person with time pressure).

Pacing

Pacing was examined by comparing network densities over time and across condition and media. Therefore, we calculated the unweighted density of a directed graph. The formula used is from Wasserman and Faust (1994):

$$\Delta = \frac{L}{g(g-1)}$$

Where L = the number of arcs in the network
g = the number of possible arcs in the network

In our data, an arc in the network exists when a transition between topics occurs. We chose to use the unweighted density because our interest is in whether or not teams were discussing different topics during a given time period. This interest is independent of how often they were changing the subject, which would have suggested considering a weighted score.

The null hypotheses tested for pacing are the same as the null hypotheses employed to examine timing, therefore the same statistical tests were conducted.

Findings for Team Communication Timing and Pacing

Network pictographs of each quartile were constructed for each type of team. The pictographs for both performance episodes are provided in Fig. 5.2.

Examining Communication Timing

Frequency Differences Over Time

Frequency Differences Over Time Results

In response to the first research question, we examine changes in topic frequency across the four quartiles. A summary of the statistical results is provided in Table 5.1 in the Cross Quartile Comparisons columns. Significant differences were identified for action, mission analysis, operational strategy (during the second performance episode only), and tactical strategy. Specifically, the numbers of messages about action significantly increased over time for all types of teams across almost every quartile, with *p*-values ranging from 0.06 to <0.0001. The only non-significant changes in the numbers of action messages exchanged occurred from Quartile 3 (Q3) to Q4 for unrestricted in person teams and distributed teams facing time pressure during the second performance episode.

For mission analysis, all types of teams in both performance episodes exchanged significantly fewer messages from Q1 to Q2 at the $p < 0.0001$ level. Additionally, the unrestricted distributed teams exchanged significantly fewer messages from Q2 to Q3 during both performance episodes ($p_{T1} < 0.0001$; $p_{T2} = 0.004$) and from Q3 to Q4 during the first performance episode ($p = 0.001$). The unrestricted in person teams also exchanged significantly fewer messages about mission analysis from Q2 to Q3 during both the first ($p = 0.04$) and the second ($p = 0.10$) performance episodes. Distributed teams under time pressure exchanged significantly fewer messages from Q2 to Q3 ($p = 0.04$) and from Q3 to Q4 ($p = 0.06$) during the first performance episode. Finally, time pressured in person teams exchanged significantly fewer mission analysis messages from Q3 to Q4 in both the first ($p = 0.007$) and the second ($p = 0.03$) performance episodes.

During the second performance episode, the numbers of messages about operational strategy significantly decreased from Q1 to Q2 for unrestricted in person ($p = 0.008$) and time pressured distributed ($p = 0.002$) teams. Overall, the mean numbers of operational strategy messages exchanged was very low, ranging from 0.04 to 2.0 messages during the first performance episode and from 0 to 3.6 during the second.

Messages about tactical strategy during the first performance episode significantly decreased from Q2 to Q3 and from Q3 to Q4 for all types of teams, with *p*-values ranging from 0.01 to 0.0007. Additionally, during the first performance

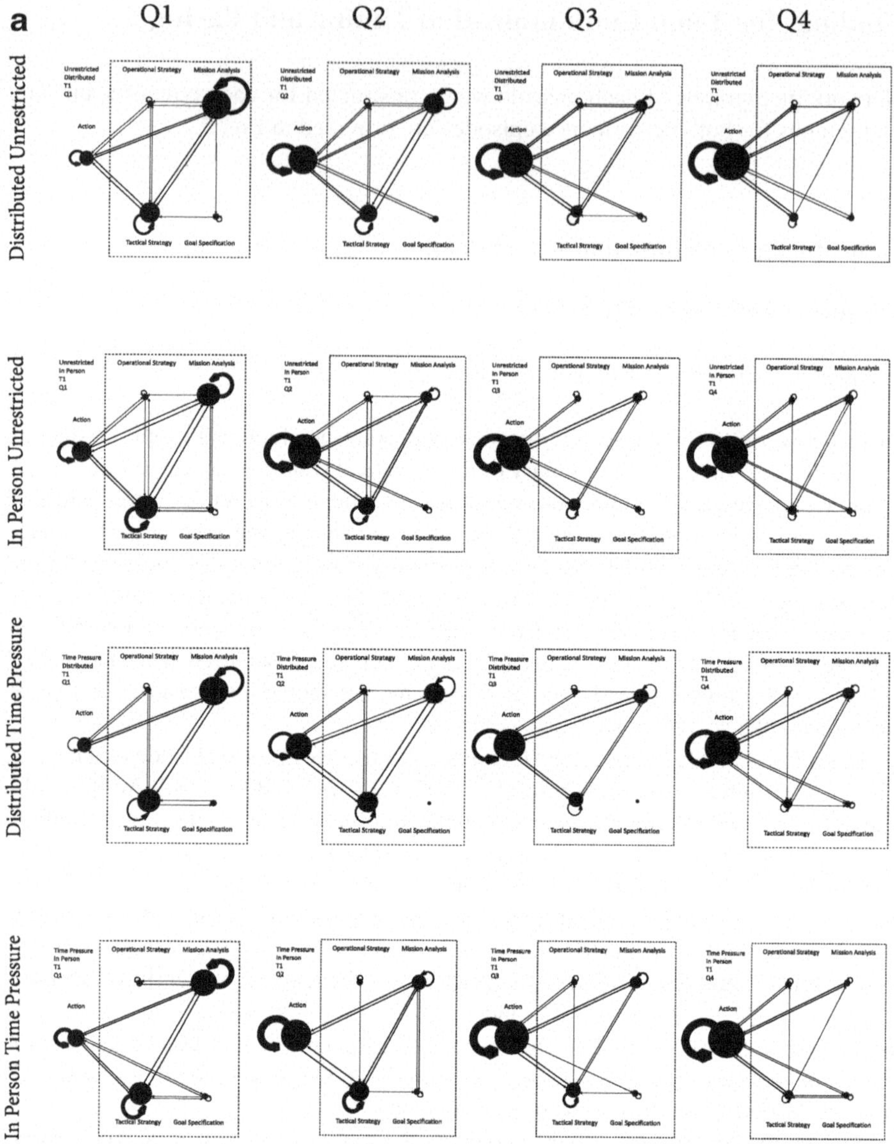

Fig. 5.2 Network pictographs for all conditions across both performance episodes. (**a**) Network pictographs for performance episode 1, (**b**) Network pictographs for performance episode 2

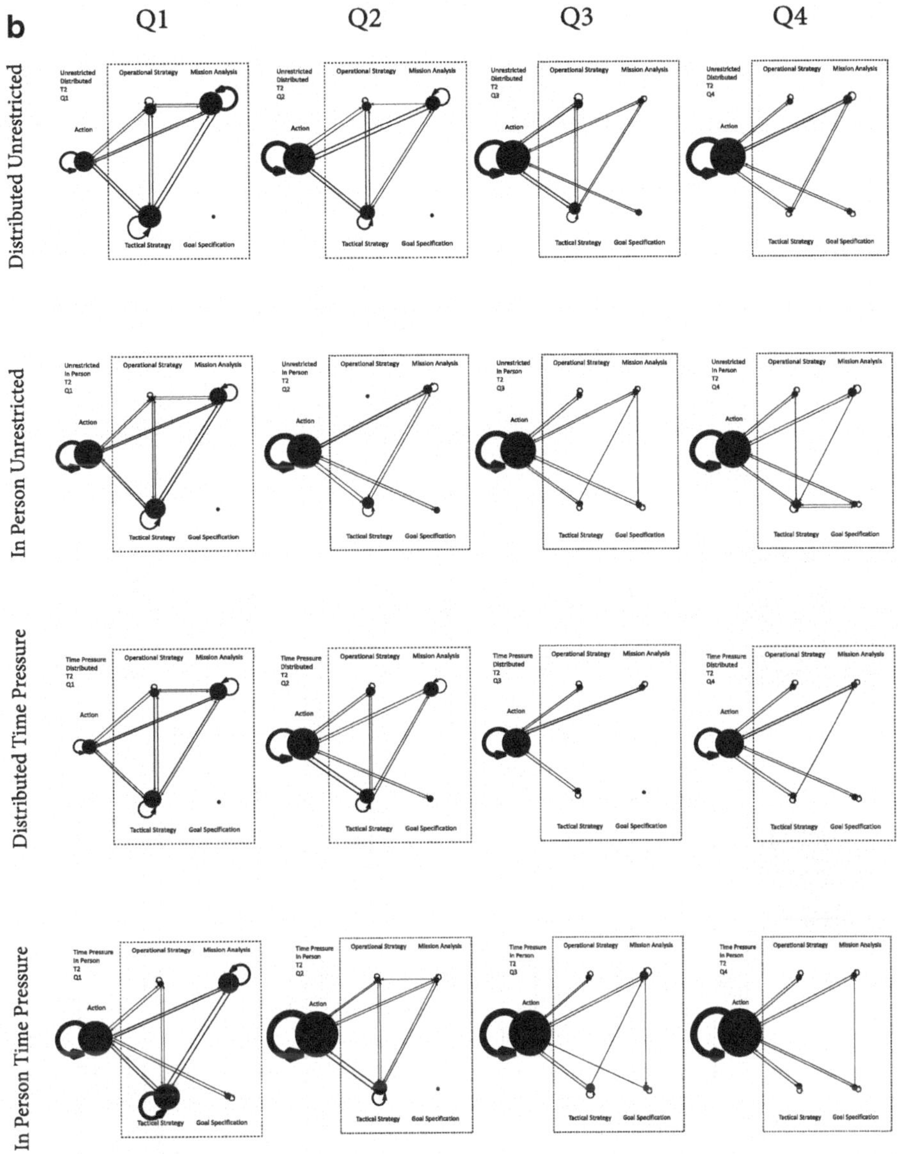

Fig. 5.2 (continued)

episode, unrestricted in person teams reduced the numbers of messages exchanged from Q1 to Q2 ($p = 0.002$). During the second performance episodes, all types of teams exchanged significantly fewer messages about tactical strategy from Q1 to Q2, with p-values ranging from 0.006 to <0.0001. The unrestricted distributed teams exchanged significantly fewer messages from Q3 to Q4 ($p = 0.04$), while the other types of teams exchanged significantly fewer messages from Q2 to Q3 (unrestricted

Table 5.1 Timing summary statistics and results

Topic	Condition	Media	Quartile	Frequency means and standard deviations performance episode 1		Frequency means and standard deviations performance episode 2		Cross quartile comparisons (p-values) (Q_n to Q_n+1)		Cross condition comparisons (p-values) Performance episode 1 in top (white) Performance episode 2 in bottom (grey)			
				M	SD	M	SD	Performance episode 1 mean comparison	Performance episode 2 mean comparison	Comparison with same quartile unrestricted distributed	Comparison with same quartile unrestricted in person	Comparison with same quartile time pressure distributed	Comparison with same quartile time pressure in person
Action	Unrestricted	Distributed n_{T1}=35 n_{T2}=35	Q1	15.1	14.4	15.4	15.6	<0.0001	<0.0001				
			Q2	35.2	28.5	34.3	23.0	<0.0001	0.002				0.006
			Q3	52.4	32.6	41.0	21.8	0.0001	0.009			0.03	0.01
			Q4	61.8	27.3	47.0	22.1					0.01	0.002
		In person n_{T1}=35 n_{T2}=28	Q1	19.0	19.5	24.8	26.5	<0.0001	<0.0001			0.05	
			Q2	54.9	44.1	41.4	35.7	0.005	0.05			0.0004	
			Q3	63.7	50.5	45.2	37.1	0.07				0.0007	
			Q4	67.3	52.0	44.9	37.2					0.002	0.01
	Time pressure	Distributed n_{T1}=30 n_{T2}=30	Q1	5.6	8.8	8.5	8.5	<0.0001	<0.0001		0.03		0.004
			Q2	17.6	14.1	22.7	13.4	0.003	0.0005		0.07		<0.0001
			Q3	25.7	16.7	26.9	10.8	<0.0001			0.07		<0.0001
			Q4	32.2	15.9	28.3	9.4			0.04			<0.0001
		In person n_{T1}=28 n_{T2}=27	Q1	24.3	32.6	37.1	33.4	<0.0001	<0.0001	0.001		<0.0001	
			Q2	65.8	48.2	66.7	38.2	0.01	0.007	0.0001	0.007	<0.0001	
			Q3	82.4	43.7	71.7	38.0	0.004	0.03	0.0003	0.005	<0.0001	
			Q4	98.1	46.5	75.2	37.4			0.0008	0.0006	<0.0001	
Goal specification	Unrestricted	Distributed	Q1	0.23	1.4	0	0						
			Q2	0.03	0.17	0	0						
			Q3	0.29	1.2	0.03	0.17						
			Q4	0.14	0.50	0.20	1.0						
		In person	Q1	0.06	0.24	0	0						
			Q2	0.14	0.69	0.03	0.19						
			Q3	0.09	0.37	0.24	1.1						
			Q4	0.37	1.2	0.44	1.0						
	Time pressure	Distributed	Q1	0.03	0.18	0	0						
			Q2	0	0	0.03	0.18						
			Q3	0	0	0	0						
			Q4	0.20	0.55	0.30	1.1						
		In person	Q1	0.32	1.1	0.18	0.94						
			Q2	0.17	0.77	0	0						
			Q3	0.11	0.42	0.18	0.94						
			Q4	0.64	1.4	0.36	0.91						
Mission analysis	Unrestricted	Distributed	Q1	34.1	18.6	18.6	11.1	<0.0001	<0.0001			0.002	0.02
			Q2	17.3	13.0	6.4	9.9	<0.0001	0.004		<0.0001	0.006	0.10
			Q3	7.1	6.2	1.8	3.7	0.001					
			Q4	2.7	4.4	1.7	3.0						
		In person	Q1	29.2	18.4	9.7	6.4	<0.0001	<0.0001	0.0002		0.07	0.0007
			Q2	6.6	6.2	2.1	3.6	0.04	0.10	0.02			
			Q3	4.1	5.0	1.0	1.9						0.07
			Q4	3.7	8.4	0.52	1.4						
	Time pressure	Distributed	Q1	18.4	11.4	9.4	6.2	<0.0001	<0.0001	<0.0001			<0.0001
			Q2	9.4	7.8	1.9	3.2	0.04		0.02			
			Q3	5.3	6.5	1.3	2.4	0.06					
			Q4	3.5	6.2	1.7	3.0						
		In person	Q1	46.8	19.8	14.8	7.4	<0.0001	<0.0001		0.09	0.07	
			Q2	11.7	9.1	1.8	3.0			0.01			
			Q3	8.6	11.0	2.5	4.2	0.007	0.03				
			Q4	2.3	3.3	0.64	1.2						

Operational strategy	Unrestricted	Distributed	Q1	1.6	3.0	3.6	4.7				0.03		0.01
			Q2	1.4	3.1	2.8	4.7				0.02	0.02	0.009
			Q3	1.3	2.8	2.1	3.6				0.04		
			Q4	2.0	3.1	1.0	2.2				0.007	0.01	0.006
		In person	Q1	0.34	1.1	0.72	1.4		0.008	0.002			
			Q2	0.23	0.55	0	0			0.0003			
			Q3	0.09	0.51	0.10	0.41			0.0007			
			Q4	0.46	1.1	0.55	1.8						
	Time pressure	Distributed	Q1	0.57	1.9	2.7	3.1		0.002		0.07		
			Q2	0.17	0.75	0.47	1.4			0.003			
			Q3	0.73	1.8	0.30	1.0			0.003			
			Q4	0.47	1.5	0.43	1.6						
		In person	Q1	0.07	0.38	0.32	0.86			0.0003		0.02	
			Q2	0.04	0.19	0.21	0.79		0.08	0.001			
			Q3	0.29	1.5	0.11	0.57			0.0008			
			Q4	0.32	1.1	0.18	0.94						
Tactical strategy	Unrestricted	Distributed	Q1	16.9	14.1	12.4	9.6		0.0001				0.01
			Q2	14.1	12.9	6.5	8.9	0.002					0.006
			Q3	6.9	5.0	5.1	7.7	0.0007	0.04				
			Q4	2.9	4.1	1.8	5.4						
		In person	Q1	25.9	29.3	12.2	11.1	0.002	0.003			0.03	
			Q2	12.6	15.0	3.8	9.4	0.01	0.09				0.002
			Q3	6.5	10.4	0.72	1.4	0.04		0.005			0.1
			Q4	4.1	8.9	2.1	7.0						
	Time pressure	Distributed	Q1	11.6	7.4	8.8	6.1		0.006				0.0007
			Q2	9.0	8.5	4.3	5.3	0.002	0.0008				<0.0001
			Q3	4.5	4.9	0.90	2.6	0.003		0.007			0.02
			Q4	1.3	2.2	0.20	0.92						
		In person	Q1	32.9	23.5	23.6	17.6		<0.0001	0.001	0.002	<0.0001	
			Q2	26.7	20.9	7.5	9.4	0.007	0.003				
			Q3	13.0	19.0	2.3	5.8	0.02					
			Q4	4.6	5.9	1.2	5.7						

in person $p = 0.09$; time pressured distributed $p = 0.0008$; time pressured in person $p = 0.003$).

No significant differences were found for goal specification. The lack of statistically significant differences in the frequency with which goal specification was discussed is not particularly surprising given the overall low numbers of messages focused on this topic. The topic frequency means range from 0 to 0.64 messages per quartile, with 8 messages being the maximum number exchanged by any team during a single quartile. In contrast, the frequency means for the other topics are routinely in double digits, particularly in the first quartile, with action ranging from 5.6 to 98.1 average messages per quartile.

Overall Patterns Observed in Frequency Differences Over Time

Differences in the frequency of communication over time were observed, thereby providing an answer to our first research question. Moreover, three interesting overall patterns are worth noting. First, the frequencies with which mission analysis, operational strategy, and tactical strategy were discussed decreased over time. Conversely, the frequency with which action was discussed increased over time. This finding aligns with the normal progression of team activity. Early in their life-cycles, teams must engage in transition processes to garner an understanding of what needs to be done and how they are going to do it. They then shift to doing the work required to meet their objectives.

Second, the frequencies with which mission analysis, operational strategy, and tactical strategy were lower in the second performance episode when compared to the frequencies in the first performance episode. The pattern was opposite for action, with higher frequencies occurring in the second performance episode. Again, this finding reflects how one would anticipate that teams behave. For the second performance episode, our teams were given the same basic task with slightly different requirements (i.e., employee staffing levels were by day in the first performance episode and by two-hour time block in the second). Therefore, they started the second performance episode with an understanding of the task and what needed to be done. Only limited conversations about transition processes were required to incorporate the change in requirements.

Third, the majority of significant changes over time occurred in the frequency with which teams discussed action, mission analysis and tactical strategy. For these topics, the frequencies were different across almost every quartile. Several notable differences can be seen in these results. For action, the frequencies with which the teams discussed it remained similar across the third and fourth quartile of the second performance episode. This finding aligns with our earlier observation that teams can engage in the work required to complete the project much more quickly in the second performance episode since they begin with a basic understanding of what is required.

Mission analysis frequencies decreased throughout the teams' performance episodes, with the most precipitous drop occurring from the first to the second quartile.

This pattern is logical since mission analysis represents discussions about the task objectives, environment, and resources. For tactical strategy, which represents how teams will complete their task, the frequencies dropped significantly from the second to the third quartile during the first performance episode (with the exception of in person teams with unlimited time). During the second performance episode, the significant change occurred from the first to the second quartile and then dropped again for the third (i.e., unrestricted in person, time pressured distributed, time pressured in person) or fourth quartile (i.e., unrestricted distributed) to two or fewer messages exchanged on average. These conversations about mission analysis and tactical strategy appear to be sequential. In other words, the teams establish what needs to be done and the parameters associated with that work in the first quartile. In the second quartile they shift their focus to how they are going to approach their task based on those established parameters.

Operational strategy was only significantly different from the first to the second quartile during the second performance episodes of distributed teams with time pressure and in person teams with unlimited time. During the other quartiles, the frequencies remained relatively low throughout the performance episodes. Overall, the frequency of operational strategy messages being exchanged is higher during the second performance episode. One possible explanation may be that more protracted discussions about how roles and responsibilities of members may only be meaningful after the team has a clear understanding about what they are charged with accomplishing and how they are going to go about accomplishing it.

Frequency Differences Across Time Constraints and Communication Media

Frequency Differences Across Time Constraints and Communication Media Results

The second and third research questions focus on differences in message frequency among teams working in the different study conditions (i.e., time constraints and communication media). These results are also reported in Table 5.1 and can be found in the Cross Condition Comparisons columns. Significant differences were found for all topics except goal specification. But, as previously described, the lack of significance associated with goal specification may be attributable to the low volume of messages about that topic. Given that the patterns of significant differences varied by topic, we describe the results for each topic separately.

With only a few exceptions, time pressured in person teams exchanged significantly more action messages than any other type of team across all quartiles, with p-values ranging from 0.01 to <0.0001. For the exceptions, no significant differences were detected; these exceptions occurred in the first performance episode during Q1 when the time pressured in person teams were compared to unrestricted distributed teams and during Q1, Q2, and Q3 when compared to unrestricted in person teams. During the second performance episode, the only non-significant difference occurred in Q1 when time pressured in person teams were compared to unrestricted in person teams. The unrestricted in person teams exchanged significantly more action messages than time pressured distributed teams across all quartiles of both performance episodes, except Q4 of the second performance episode, with *p*-values ranging from 0.05 to 0.0004 in the first performance episode and from 0.07 to 0.03 in second performance episode. The unrestricted distributed teams exchanged significantly more action messages than time pressured distributed teams during both Q3 ($p = 0.03$) and Q4 ($p = 0.01$) of the first performance episode, as well as in Q4 ($p = 0.04$) of the second performance episode.

For mission analysis, the unrestricted distributed teams exchanged significantly more messages during first performance episode than unrestricted in person teams during Q2 ($p < 0.0001$) and time pressured distributed teams during Q1 ($p = 0.002$) and Q2 ($p = 0.006$). During the second performance episode, the unrestricted distributed teams exchanged significantly more mission analysis messages than unrestricted in person teams during Q1 ($p = 0.0002$) and Q2 ($p = 0.02$), time pressured distributed teams during Q1 ($p < 0.0001$) and Q2 ($p = 0.02$), and time pressured in person teams during Q2 ($p = 0.01$). In the first performance episode, the time pressured in person teams exchanged significantly more messages than unrestricted distributed teams during Q1 ($p = 0.02$) and Q2 ($p = 0.10$), unrestricted in person teams during Q1 ($p = 0.0007$) and Q3 ($p = 0.07$), and time pressured distributed teams in Q1 ($p < 0.0001$). The time pressured in person teams also exchanged significantly more mission analysis messages than time pressured distributed teams ($p = 0.07$) and unrestricted in person teams ($p = 0.09$) during Q1 of the second performance episode. Finally, the unrestricted in person teams exchanged significantly more mission analysis messages than time pressured distributed teams during Q1 ($p = 0.07$) of the first performance episode.

The significant results for operational strategy during the first performance episode occurred only in comparisons with unrestricted distributed teams. Specifically, these teams exchange significantly more messages than unrestricted in person teams across all quartiles ($p_{Q1} = 0.03$; $p_{Q2} = 0.02$; $p_{Q3} = 0.04$; $p_{Q4} = 0.007$), time pressured distributed teams during Q1 ($p = 0.02$) and Q4 ($p = 0.01$), and time pressured in person teams in Q1 ($p = 0.01$), Q2 ($p = 0.009$), and Q4 ($p = 0.006$). During the second performance episode, the unrestricted distributed teams exchanged significantly more operational strategy messages than unrestricted in person teams in Q1 ($p = 0.002$), Q2 ($p = 0.0003$), and Q3 ($p = 0.0007$), time pressured distributed teams in Q2 ($p = 0.003$) and Q3 ($p = 0.003$), and time pressured in person teams in Q1 ($p = 0.0003$), Q2 ($p = 0.001$), and Q3 ($p = 0.0008$). Finally, the time pressured distributed teams exchanged significantly more messages during Q1 than the unrestricted in person teams ($p = 0.07$) and time pressured in person teams ($p = 0.02$) during the second performance episode.

For tactical strategy, the time pressured in person teams exchanged significantly more messages during the first performance episode than unrestricted distributed teams in Q1 ($p = 0.01$) and Q2 ($p = 0.006$), unrestricted in person teams in Q2 ($p = 0.002$) and Q3 ($p = 0.1$), and time pressured distribute teams in Q1 ($p = 0.0007$), Q2 ($p < 0.0001$), and Q3 ($p = 0.02$). Also during the first performance episode, the unrestricted in person teams exchanged more messages during Q1 than the time pressured distributed teams ($p = 0.03$). Shifting to the second performance episode, the time pressured in person teams exchanged significantly more messages during Q1 than unrestricted distributed teams ($p = 0.001$), unrestricted in person teams ($p = 0.002$) and time pressured distributed teams ($p < 0.0001$). For the second performance episode, the unrestricted distributed teams exchanged significantly more tactical strategy messages during Q3 than unrestricted in person teams ($p = 0.005$) and time pressured distributed teams ($p = 0.007$).

Overall Patterns Observed in Frequency Differences Across Time Constraints and Communication Media

When examining differences in the frequency of communication across time constraints and communication media in response to our second and third research questions, respectively, differences were observed. Indeed, several interesting overall patterns are worth noting about the way in which messages about action, mission analysis, operational strategy, and tactical strategy were exchanged.

For messages about action, the patterns of results were similar for both performance episodes. First, the in person teams talked significantly more than the distributed teams. This result may be attributable to the relative ease of conversing in person versus exchanging typed messages via a computer interface. Second, the volume of messages for distributed teams was higher for those teams with unlimited time versus those with time pressure. The teams with unlimited time may have had the luxury of exchanging messages about the work they were doing, whereas those teams facing time pressure were more focused on accomplishing the work without taking time to comment about it via a medium that was not conducive to easy conversation. This relationship was opposite for in person teams, where those teams with time pressure exchanged significantly more messages than those with unlimited time, particularly in the second performance episode. Indeed, the in person teams facing time pressure spoke significantly more about action than any other types of team in both performance episodes. This need to actively converse about the work being done may represent an effort to keep the team focused on what they needed to accomplish in the relatively short time available to accomplish it.

Two interesting patterns of significant differences can be observed for mission analysis. First, while the in person teams with time pressure exchanged the most messages about mission analysis during the first quartile of the first performance episode (significantly more than all three other types of teams), they achieved parity with the other types of teams by the second quartile. The unrestricted teams, particularly those in the distributed condition, continued to have significantly more exchanges through the second quartile. Second, the distributed teams with unlimited time exchanged significantly more messages about mission analysis during the first two quartiles of the second performance episode. Visual cues may help in person teams demonstrate an ability to comprehend their task objectives and other parameters much more quickly than distributed teams. The information sheets distributed to all team members were different colors for each team member and the text in the final paragraph was different (all other text was identical). These visual cues may have helped the in person teams realize that they needed to clarify their parameters. Alternatively, the distributed teams may not have realized the need for this clarification until information not available to all team members was communicated explicitly.

The distributed teams with unlimited time exchange significantly more messages about operational strategy than any of the other types of teams. This pattern can be observed throughout all quartiles of the first performance episode and in the first three quartiles of the second performance episode. During the second performance episode, the distributed teams facing time pressure started their time together exchanging significantly more messages about operational strategy than either of the in person teams. Similar to mission analysis, the visual cues available to in person teams about which team members were undertaking which activities were not available to distributed teams. Therefore, distributed teams needed to articulate member roles and responsibilities because they could not see what each other was doing.

Finally, the results associated with tactical strategy focus on in person teams with time pressure and distributed teams with unlimited time. First, the in person teams facing time pressure exchanged significantly more messages about tactical strategy during the first three quartiles of the first performance episode than the other types of teams. In the second performance episode, they continued to exchange significantly more messages about tactical strategy during the first quartile. The limited time available to the time pressured teams in conjunction with the ease of communicating may be the impetus for these strategic conversations about how the team was going to accomplish its task. Second, the distributed teams with unlimited time exchanged significantly more messages about tactical strategy during the third quartile of the second performance episode than either the distributed teams facing time pressure or the in person teams with unlimited time. The need for distributed teams with unlimited time to continue discussing tactical strategy more actively than other teams into the third quartile may be attributable to their continued conversations about mission analysis into the first quartile of the second performance episode. If these teams were still not clear about what they were tasked with accomplishing, their inability to determine who was responsible for what action is understandable.

Investigating Communication Pacing

Complexity Differences Over Time and Across Time Constraints and Communication Media

Complexity Differences Over Time and Across Time Constraints and Communication Media Results

The fourth research question focuses on changes in the complexity of team conversations across the four quartiles. The statistical results can be found in Table 5.2 in the Cross Quartile Comparisons columns. During the first performance episode,

Table 5.2 Pacing summary statistics and results

Condition	Media	Quartile	Density means and standard deviations performance episode 1		Density means and standard d eviations performance episode 2		Cross quartile c omparisons (*p*-values) (Q_n to Q_n+1)			Cross condition comparisons (*p*-values) performance episode 1 in top (white) performance episode 2 in bottom (grey)		
			M	*SD*	*M*	*SD*	Performance episode 1 density comparison	Performance episode 2 density comparison	Comparison with same quartile unrestricted distributed	Comparison with same quartile unrestricted in person	Comparison with same quartile time pressure distributed	Comparison with same quartile time pressure in person
Unrestricted	Distributed n_{T1}=35 n_{T2}=35	Q1	0.013	0.0053	0.015	0.0059		0.0002			0.005	
		Q2	0.013	0.0050	0.009	0.0058		0.09			0.002	
		Q3	0.012	0.0047	0.007	0.0043	0.0002	0.001		<0.0001	0.0004	0.01
		Q4	0.008	0.0053	0.004	0.0044						
	In person n_{T1}=35 n_{T2}=28	Q1	0.012	0.0050	0.012	0.0061		<0.0001			0.1	
		Q2	0.011	0.0051	0.004	0.0046	<0.0001		0.0009			
		Q3	0.007	0.0049	0.004	0.0037			0.002			
		Q4	0.007	0.0052	0.004	0.0047						
Time pressure	Distributed n_{T1}=30 n_{T2}=30	Q1	0.009	0.0057	0.012	0.0057		<0.0001				
		Q2	0.009	0.0048	0.005	0.0041		0.02	0.008			
		Q3	0.007	0.0040	0.003	0.0033	0.06		0.0002			
		Q4	0.006	0.0048	0.003	0.0036						
	In person n_{T1}=28 n_{T2}=27	Q1	0.011	0.0050	0.011	0.0059		0.0006	0.05			
		Q2	0.011	0.0053	0.005	0.0049	0.07	0.05	0.01			
		Q3	0.009	0.0047	0.003	0.0038			0.0006			
		Q4	0.007	0.0040	0.002	0.0030						

team conversations significantly decreased in complexity from Q2 to Q3 for unrestricted in person ($p < 0.0001$) and time pressured in person ($p = 0.07$) teams. Significant decreases in the complexity of the conversations were also identified from Q3 to Q4 for unrestricted distributed ($p = 0.0002$) and time pressured distributed ($p = 0.06$) teams. During the second performance episode, the complexity of conversations also decreased significantly over time for unrestricted distributed teams ($p_{Q1\text{–}Q2} = 0.0002$; $p_{Q2\text{–}Q3} = 0.09$; $p_{Q3\text{–}Q4} = 0.001$), unrestricted in person teams ($p_{Q1\text{–}Q2} < 0.0001$), time pressured distributed teams ($p_{Q1\text{–}Q2} < 0.0001$; $p_{Q2\text{–}Q3} = 0.02$), and time pressured in person teams ($p_{Q1\text{–}Q2} = 0.0006$; $p_{Q2\text{–}Q3} = 0.05$).

Differences in the complexity of team conversations among teams working in the different study conditions (i.e., time constraints and communication media) are examined in response to the fifth and sixth research questions. We discuss all of these results together as the differences over time are different across the team types. The statistical results can be found in Table 5.2 in the Cross Condition Comparisons columns. With the exception of unrestricted in person teams having significantly more complex conversations than time pressured distributed teams during the first quartile of the first performance episode ($p = 0.1$), all other significant differences in both performance episodes indicated that the unrestricted distributed teams had significantly more complex conversations than other types of teams. Specifically, during the first performance episode, unrestricted distributed teams had more complex conversations than unrestricted in person teams in Q3 ($p < 0.0001$), time pressure distributed teams in Q1 ($p = 0.005$), Q2 ($p = 0.002$), and Q3 ($p = 0.0004$), and time pressured in person teams in Q3 ($p = 0.01$). In the second performance episode, unrestricted distributed teams were significantly different than unrestricted in person teams in Q2 ($p = 0.0009$) and Q3 ($p = 0.002$), time pressured distributed teams in Q2 ($p = 0.008$) and Q3 ($p = 0.0002$), and time pressured distributed teams in Q1 ($p = 0.05$), Q2 ($p = 0.01$), and Q3 ($p = 0.0006$).

Overall Patterns Observed in Complexity Differences Over Time and Across Time Constraints and Communication Media

Our results confirm that differences in communication complexity occur in team conversations over time (RQ5.4), across time constraints (RQ5.5), and across communication media (RQ5.6). Several interesting overall patterns were observed in the results that may help us better understand the pacing of team communication.

During the first performance episode, distributed teams conversed about many topics throughout the first three quartiles. The in person teams became more focused starting in the third quartile. As reported in the communication timing results earlier in this chapter, the focus shifts to action in the later quartiles. This finding suggests that the distributed teams are iterating on topics associated with the transition processes (i.e., goal specification, mission analysis, operational strategy, and tactical strategy) for three quarters of their time together versus in person teams who shift more completely to action after half of their messages have been exchanged. Communication via the limited bandwidth available to distributed teams is more disjointed and parsimonious within the chat box than conversing in person. This limitation may explain why distributed teams require more time to streamline their activities.

During the second performance episode, the pattern of distributed teams requiring more time discussing the transition processes than in person teams continued. In person teams significantly reduced the complexity of their communication in the second quartile and maintained a lower level of complexity for the remaining quartiles. The time pressure teams exhibited a significant reduction in complexity from the second to the third quartile, but the drop in density score was much less dramatic than their significant drop from the first to the second quartile. In other words, all team types made the most significant reductions in communication complexity from the first to the second quartile.

Across both performance episodes, the distributed teams with unlimited time had the most complex communication patterns in every quartile when compared to the other team types. This type of team may face the most challenges. They have a difficult communication medium, which delays their abilities to conclude conversations about transition processes. Simultaneously, they have unlimited time to complete their tasks so they have no urgency to reach a point where they can move more completely into action. As an interesting comparison, distributed teams with time pressure had complexity patterns that were not significantly different than those of in person teams (with the exception of the first quartile of the first performance episode where the density score for in person teams with unlimited time was marginally significantly greater than the density score for distributed teams with time pressure at the $p = 0.10$ level). This similarity suggests that distributed teams with time pressure may have to move forward with incomplete information in spite of their limited ability to communicate via chat box.

Theoretical Implications and Future Research

Our results provide confirmation of the theoretical frameworks underlying this research study. First, Marks and colleagues (2001: 360) posit that "teams are actively engaged in different types of taskwork at different phases of task accomplishment." Indeed, they forward their recurring phase model of team processes because the cross-sectional and/or unidimensional perspectives do not adequately portray the complexity of team processes. Overwhelmingly, our results suggest that teams across all conditions shift from transition processes toward more action during both performance episodes, require less emphasis on transition processes during the second performance episode, and decrease the complexity of their conversations over time. Thus, these findings confirm that team processes change over time to meet the varying demands of their task.

Second, Bartunek and Woodman (2015) identify timing and pacing as two critical temporal dimensions of change. Time represents "the need to act" and pacing encompasses the speed that teams need "especially at the beginning … to generate enough momentum …" to complete their assigned task (Bartunek and Woodman 2015, 170). Our results provide insight into timing by identifying when certain processes may be used by teams. For example, the frequency with which the majority of teams discussed mission analysis peaked during the first quartile and then the frequency reduced or remained the same across subsequent quartiles, whereas tactical strategy was prominent throughout the first two quartiles before the frequency significantly reduced in the third and fourth quartiles. The results also depict pacing in team communication. Specifically, teams engage in substantially more complex conversations as they orient themselves to what they have been tasked to accomplish and determining how they are going to accomplish it, particularly during the second quartile. For example, all types of teams experience significant reductions in the numbers of transitions among topics from the first to the second quartile. These results, therefore, provide evidentiary support for applying at least the timing and pacing temporal dimensions within the team domain.

Our research also extends Gersick's (1988, 1989) punctuated equilibrium model of group development. In her research, Gersick identified a phase transition at the midpoint of team activity where team members significantly changed the manner in which they were collaborating no matter how long the teams' lifecycles were. In our results, we identified significant changes across four quartiles (versus Gersick's halves), particularly in the teams' action, mission analysis, and tactical strategy processes. These shifts across quartiles suggest that researchers may benefit from including more time periods in their longitudinal examinations of team behavior.

In this study, we focus on the various transition processes teams employ to organize themselves for action. In addition to these transition processes, Marks and colleagues (2001) delineate four action processes (i.e., monitoring progress toward goals, systems monitoring, team monitoring and backup behavior, and coordination activities). Team science may benefit from an in depth examination of the specific action processes to identify when and at pace they are needed to facilitate team activity.

Methodological Contributions and Future Research

Our findings in this chapter underscore how fruitful relatively simple visualization and statistical approaches can be as an early step toward investigating complex, dynamic data. We built upon the work of Olson and colleagues (1992) by providing visualizations depicting how frequently team members discuss specific topics and the transitions among those topics. Moreover, we extend their approach by breaking the conversations into quartiles and applying social network analysis to quantify transition frequencies via density scores. Our visualizations and density scores, in particular, help to capture the changing complexities of the teams' conversations through their shifts and plateaus over time.

Future researchers interested in longitudinal changes in team processes may want to consider employing relatively simple visualizations and statistical comparisons. The oft employed aggregated representations of team processes mask the rich nature of collaboration. As our findings in this chapter have shown, however, even dividing aggregated data into quartiles resulted in the identification of interesting communication patterns over time.

Managerial Minutes

Distributed teams need a communication facilitator.
If your team members are distributed, they may struggle with forward progress. These struggles are particularly difficult for teams with minimal time pressure. Under these conditions, you may need a strong facilitator to help the team reach consensus and make decisions.

Set the team on the right track by starting with conversations about transition processes.
Teams need to focus on understanding what they are being asked to do, figure out how they are going to do it, and then assign the work. Too many teams start with assignments and then realize much later that everyone is working at cross purposes. Setting the appropriate agenda for team meetings may be one way to help ensure that teams are progressing through these topics systematically.

Appendix: How to Create and Compare Network Pictographs

In Chap. 5, we examined team communication timing and pacing using network pictographs and statistical comparisons. In this appendix, we describe how we conducted these analyses and provide the results using the example dataset introduced in the Chap. 3 Appendix.

All statistical analyses were conducted in SAS 9.4. Network pictographs were constructed in PowerPoint.

Data Preparation

To conduct these analyses only topic data were used. The relevant data strings for the example dataset can be found in the bottom of Table 3.3 in the Chap. 3 Appendix.

Timing

To examine the timing of communication, we employed frequency counts of topics. We use Team 1 from the example dataset to demonstrate how we calculated the frequencies. Team 1's communication string (from Table 5.3) is:

10, 10, 10, 20, 20, 20, 20, 30, 30, 30, 10, 10, 10, 40, 40, 40, 40, 40, 40, 40

The total number of messages exchanged by Team 1 is 20. When we break down the total number of exchanges by topic, the frequency count for topic 10 is 6, for topic 20 is 4, for topic 30 is 3, and for topic 40 is 7. The topic frequency counts for all teams in the example dataset are given in Table 5.3.

We divided the data strings into halves by dividing the total number of exchanges by 2. In Chap. 5, we divided the data strings into quartiles, but those communication strings are much longer than the strings in the example dataset used in this appendix. These divisions facilitate identifying any differences over time.

Two datasets were created in SAS 9.4, one for condition 1 and another for condition 2. Note that four datasets were created in Chap. 5 for the four conditions in which teams collaborated (i.e., unrestricted distributed, unrestricted in person, time pressured distributed, and time pressured in person teams).

Pacing

To examine the pacing of communication and create the network pictographs, data strings were converted into strings of t+1 transitions. For example, since the first two messages are coded as topic 10, the first t+1 transition is 1010. The t+1 transitions are:

1010, 1010, 1020, 2020, 2020, 2020, 2030, 3030, 3030, 3010, 1010, 1010, 1040, 4040, 4040, 4040, 4040, 4040, 4040

The transition frequency counts for the example dataset are given in Table 5.3. Frequency counts of t+1 transitions for each half were calculated (these counts were by quartile in Chap. 5). Datasets were created in SAS 9.4 for each type of team.

Network Pictographs

Network pictographs were constructed by creating a template within PowerPoint, where circles represented the topics and lines represented the t+1 transitions. Next, for each half within each condition, the average frequency of messages about specific topics and t+1 transitions between all possible topic pairings (e.g., 10 to 10, 30 to 20) were calculated and scaled. A scaling algorithm was established that balanced the limitations within PowerPoint with the need to easily visualize differences. The PowerPoint template was then used to create network pictographs, one for each half of each condition for a total of four diagrams, which are presented in Fig. 5.3. The

Table 5.3 Topic and t+1 transition frequency counts for the example dataset

		Topic Frequencies				Transition Frequencies															
Team	Condition	Topic 10	Topic 20	Topic 30	Topic 40	1010 Transitions	1020 Transitions	1030 Transitions	1040 Transitions	2010 Transitions	2020 Transitions	2030 Transitions	2040 Transitions	3010 Transitions	3020 Transitions	3030 Transitions	3040 Transitions	4010 Transitions	4020 Transitions	4030 Transitions	4040 Transitions
Team1	1	6	4	3	7	4	1	0	1	0	3	1	0	1	0	2	0	0	0	0	6
Team2	1	5	7	5	5	3	1	1	0	0	5	1	1	1	1	3	0	0	0	0	4
Team3	1	9	4	5	6	6	1	1	1	1	2	1	0	2	0	3	0	0	0	0	5
Team4	1	6	5	5	6	4	0	1	1	0	4	1	0	0	1	3	1	1	0	0	4
Team5	1	6	5	5	8	4	0	1	1	1	3	0	1	1	0	3	1	0	1	1	5
Team6	2	7	3	6	6	5	0	0	2	0	2	1	0	2	0	4	0	0	1	0	4
Team7	2	10	2	3	5	8	0	1	1	0	1	0	1	0	0	2	1	1	1	0	2
Team8	2	6	3	7	6	4	1	0	1	0	2	1	0	2	0	5	0	0	0	1	4
Team9	2	9	4	3	6	6	1	1	0	0	2	0	2	0	0	2	1	2	1	0	3
Team10	2	9	6	3	4	6	2	0	1	0	4	1	1	1	0	2	0	1	0	0	2

relative circle sizes and line weights can be used to compare how frequently a topic was discussed and a transition occurred, respectively. For example, in the first half of condition 1, topics 10, 20, and 30 were discussed far more frequently than topic 40 and the transitions within topics 10, 20, and 30 (i.e., 10 to 10, 20 to 20, and 30 to 30) occurred more frequently than the transitions between topics.

Data Usage

The means for the entire sample, as well as the means by condition and half are given in Table 5.4.

Timing

To assess changes in topic frequency over time, repeated measures analysis of variance (ANOVA) was run for each type of team separately. The significant p-values up to $p < 0.1$ level for the example dataset are provided in Table 5.4 in the column Comparisons Across Halves. The results show, for example, that in condition 1, the mean frequency with which teams discuss topic 10 drops from 4.6 in the first half to 1.8 in the second; and this drop is significant at the $p = 0.08$ level. Topic 10 is significantly different from the first to the second half of condition 2 ($p = 0.03$) as well.

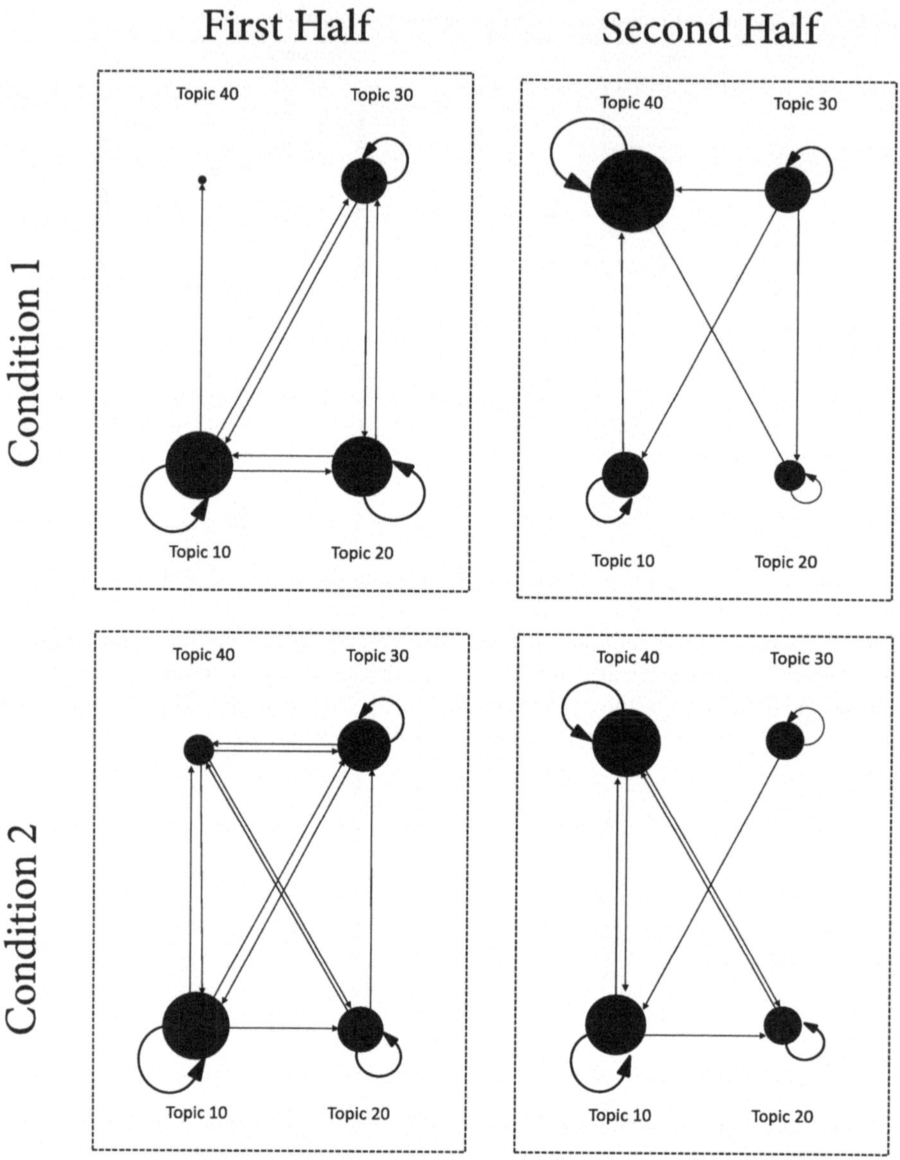

Fig. 5.3 Network pictographs for example dataset

Topic 20, however, is only significantly different from the first to the second half for teams in condition 1 ($p = 0.007$).

Topic frequencies across types of teams were also analyzed. A univariate ANOVA was used to obtain the results, which are reported in Table 5.4 in the column Comparisons Across Conditions. The results suggest, for example, that topic 10 is

Table 5.4 Sample statistics and results for teams in the example dataset

Topic	Total sample *M*	Total sample *SD*	Condition	Half	*M*	*SD*	Comparison across halves significant *p*-values	Comparison across conditions significant *p*-values
Topic 10	7.3	1.8	Condition 1	First	4.6	1.5	0.08	
				Second	1.8	1.6		0.05
			Condition 2	First	4.6	1.1	0.03	
				Second	3.6	0.5		
Topic 20	4.3	1.5	Condition 1	First	4.2	0.4	0.007	0.01
				Second	0.8	1.3		
			Condition 2	First	2.0	1.4		
				Second	1.6	1.5		
Topic 30	4.5	1.4	Condition 1	First	2.4	0.5		0.04
				Second	2.2	1.3		
			Condition 2	First	3.2	0.4	0.03	
				Second	1.2	1.6		
Topic 40	5.9	1.1	Condition 1	First	0	0	0.0002	0.01
				Second	6.4	1.1		0.01
			Condition 2	First	1.0	0.7	0.001	
				Second	4.4	0.5		
Density	0.043	0.009	Condition 1	First	0.029	0.0063	0.08	
				Second	0.018	0.0086		
			Condition 2	First	0.029	0.0063		
				Second	0.021	0.0063		

$N_{all} = 10$; $N_{condition1} = N_{condition2} = 5$

not significantly different across conditions during the first half; during the second half, however, teams in condition 2 exchange topic 10 messages significantly more frequently than those teams in condition 1 ($p = 0.05$). Specifically, during the second half of their time collaborating, condition 1 teams exchanged an average of 1.8 topic 10 messages, whereas condition 2 teams exchanged an average of 3.6 messages.

In Chap. 5, the data includes four types of teams that completed the laboratory task in an unrestricted or time pressured condition (i.e., condition) and either via distributed or in person (i.e., communication media). These additional dimensions required running a full factorial ANOVA to test for differences across study conditions.

Pacing

Communication pacing was operationalized as the unweighted density of a directed graph, which was calculated for each half and across conditions. In Chap. 5, the calculations were done by quartile and across the four study conditions. The formula used is from Wasserman and Faust (1994):

$$\Delta = \frac{L}{g(g-1)}$$

Where L = the number of arcs in the network
g = the number of possible arcs in the network

In our data, an arc in the network exists when a transition between topics occurs. Our interest in whether or not teams discussed different topics during a given time period was the rationale for employing an unweighted density score. If our interest was dependent upon how often they changed the subject, a weighted density score would have been more appropriate.

The results of our examination of pacing are reported in the bottom row of Table 5.4. Density changes over time and across conditions were assessed via repeated measures ANOVA and full factorial ANOVA, respectively. The procedures followed were the same as described above for timing. The only significant difference in pacing in the example dataset was from the first half to the second half in teams assigned to condition 1 ($p = 0.08$). Specifically, the density scores dropped from 0.029 to 0.018 across the two halves, suggesting that the conversations of these teams became significantly more streamlined over time.

Chapter 6
Polyphony in Team Communication

When teams are convened, they have many issues to discuss: what have we been asked to do, what should our deliverable look like, how should we make it, etc. Some of these conversations occur all at once; while, others require shifting among topics until enough information is available to connect the dots and make key decisions that reflect all the information available. Understanding how these conversations transpire, including when topics are discussed and in what order, may help teams perform effectively. In this chapter, we consider communication polyphony, which results in identifying the relationships among the points in time when teams engage in conversations about transition and action processes. The findings from these analyses highlight the order in which conversations about transition processes are completed, when action is initiated, how much time elapses between these points of interest, how often teams cycle through various topics, and any relationships these points of interest have with team performance. Differences attributable to the conditions under which the teams collaborate are also examined. Insights about the polyphony of communication over time will help team leaders maintain forward progress by, for example, identifying when to nudge their teams toward consensus, suggesting they revisit a topic, or reminding them of relevant decisions they have already made.

Introduction to Team Communication Polyphony

Our examination of team communication now shifts to polyphony, which we examine by using event history analysis. Polyphony refers to the alignment and overlap among activities, including any dependencies among them (Bartunek and Woodman, 2015). In our study, those activities are the topics discussed by team members. This focus provides insights into how teams achieve their outcomes as a function of the manner in which topics relate to one another and the amount of time that elapses between topic

S. McComb, D. Kennedy, *Computational Methods to Examine Team Communication*, Computational Social Sciences,
https://doi.org/10.1007/978-3-030-36159-4_6

discussions. Additionally, we identify any differences in polyphony attributable to the conditions under which the teams are working (i.e., distributed versus in person; unrestricted versus time pressured) and how they may relate to team performance.

Mitchell and James (2001) propose that time lags are critical to the nature of organizations, yet more needs to be known about the lag between when one event occurs and when another event or effect follows. Event history analysis is one approach to address this need. Event histories are comprised of longitudinal data about the timing of events of interest and potential explanatory variables. These histories, however, include censoring (i.e., the events may not occur during the study period) and time-varying explanatory variables (i.e., variables that change value during the study period) that pose serious issues with the use of standard statistical methods (Allison 1984). Event history analysis is comprised of a set of related methods that have been devised to handle these eventualities in event history data. Event history analysis was employed in only one study of longitudinal team communication included in our literature review (see Chap. 2). McComb and colleagues (2010) used the proportional hazards model associated with event history analysis to examine the temporal interdependencies among discussions of different mental model content. The results demonstrate that polyphony associated with content discussions (1) differed depending on the communication media used for teamwork and (2) impacted team performance.

Following the example of McComb and colleagues (2010), we similarly apply event history analysis to examine the temporal interdependences among transition and action processes discussed by team members. We construct an event history by identifying points of interest in the communication strings. Specifically, we identify the points in time when teams had completed meaningful conversations about their transition processes and when they initiated action. We label these points in time topic shift points. Our results offer insights about *when* and *if* these topic shift points occur.

Our investigation of longitudinal team communication also extends the approach of McComb and colleagues (2010) to include the intensity of polyphony across the performance episodes, where intensity refers to the number of times a pattern repeats and is examined by using Andersen-Gill intensity models. For this analysis, we focus our attention on mission analysis and tactical strategy because many researchers have suggested that these types of activities are likely to surface early in collaboration as teams work to determine what they have been tasked with accomplishing and how they are going to accomplish it (DeChurch and Haas 2008; Mathieu and Rapp 2009). Additionally, these extant suggestions align with our Chap. 5 results, where mission analysis and tactical strategy are the most prevalently discussed transition processes. In sum, we use the Andersen-Gill intensity model to examine how often topic shift points occur (i.e., how intense these topics are) and what triggers revisiting mission analysis and tactical strategy.

As discussed in previous chapters, the media available for communication (Daft and Lengel 1986) and the time available to complete the task (David and Schraagen 2018; Davis et al. 2016; Driskell et al. 1999; Ellis 2006; Kelly and McGrath 1985) may impact how teams exchange information. We anticipate that these conditions may, therefore, require variations in the patterns of polyphony exhibited by teams. Our analyses also include team performance, because variations in patterns of polyphony may impact a team's ability to succeed (McComb et al. 2010).

Research Questions

Two sets of research questions are posed to examine polyphony exhibited by teams. First, we focus on interrelationships among topic shift points (i.e., the point at which a meaningful conversation about a specific topic is completed or action is initiated) by answering the research questions:

RQ6.1. Do topic shift points interrelate to create polyphonic patterns over time?
RQ6.2. Do polyphonic patterns differ depending on the time teams have available (i.e., unrestricted time versus time pressured)?
RQ6.3. Do polyphonic patterns differ depending on the communication media available to teams (i.e., distributed versus in person)?
RQ6.4. Do polyphonic patterns relate to team performance?

Second, we examine the repetition in team conversations by focusing on how frequently teams revisit topics and answering the research questions:

RQ6.5. Do teams repetitively iterate through meaningful conversations about mission analysis and tactical strategy over time?
RQ6.6. What triggers these repetitive iterations to create polyphonic patterns over time?
RQ6.7. Do repetitive iterations within the polyphonic patterns differ depending on the time teams have available?
RQ6.8. Do repetitive iterations within the polyphonic patterns differ depending on the communication media available to teams?
RQ6.9. Do repetitive iterations within the polyphonic patterns relate to team performance?

Methodology

Methodological Approaches: Proportional Hazards and Andersen-Gill Intensity Models

Event history analysis is the translation of survival analysis to the social sciences. Survival analysis was designed to examine the occurrence of events in longitudinal data structures (Allison 1995). The benefit of using survival analysis as opposed to using other conventional statistical methods comes from its ability to deal with certain data features. First, survival analysis allows for special treatment of *censored data* such that data can be included even if the event(s) does not occur during the study period (right censored). In our examination of team communication, right-censored data may occur when teams do not complete meaningful conversations about a particular transition process during a performance episode. Second, *time-*

varying covariates, those covariates that may change during the study period, are appropriately modeled. For example, an outcome may depend on certain covariates that remain unchanged over time (e.g., team size, communication medium) whereas other potential covariates (e.g., conversation shift points) may occur at any time during a performance episode. Furthermore, Allison (1995) suggests that the treatment of censored and time-varying covariates by other statistical approaches (e.g., logistic regression), while possible, require decreasing the information inherent in the dataset in order to function.

Many analytical approaches comprise event history analysis (Allison 1984). We employ two different approaches to address our research questions. For RQ6.1 through RQ6.4, we use the proportional hazards model to examine when initial conversation shift points occur and any interrelationships among those points. For RQ6.5 through RQ6.9, we use the Andersen-Gill intensity model, which is specifically designed to handle recurrent events. The results provide insights about the polyphony of team communication, including both the interrelationships among initial conversations about topics and any recurrence of conversations about mission analysis and tactical strategy over time.

Data Specification

Proportional Hazards Analysis

To employ the proportional hazards model, we examine communication strings for patterns that indicate topic shift points and any order among them. For transition processes, the pattern represents a period of focused conversation about the specific topic followed by messages about any other topics for a substantial period of time. As described in Chap. 3, the pattern occurs when three or more consecutive messages about a topic are exchanged and followed by at least 25 messages about other topics. Once the pattern is found, the topic shift point is designated to be the position in the communication string of the last message about the focal topic. We then scale the topic shift points to address variation in communication string lengths across teams. For example, if we identify a tactical strategy shift point at message 220 out of a communication string length of 249 messages, we designate the shift point for tactical strategy to be $220/249 = 0.88$. For messages with the topic coded as action, we identify the first time the team shifts to communicate about action processes. We use the position in the communication string where the topic begins and scale it by the communication string length. Thus, the shift point for action when found at position 14 in a string length of 249 is $14/249 = 0.06$. The dataset used to address RQ6.1 through RQ6.4 included (1) the scaled shift points for all transition processes and action and (2) dynamic dummy variables (DDV) that designate whether the first topic shift of the transition processes or action occurred before the first shift point of the dependent variable.

Andersen-Gill Intensity Analysis

We followed an analogous approach to create the dataset required to answer research questions 6.5 through 6.9. To prepare our data for analyses, we required three types of variables. First, we included the first topic shift points, which are the same variables used for the proportional hazards analyses. Second, we counted how many transition topic shift points occur for each transition process. Third, we recorded information about when every mission analysis and tactical strategy shift point occurred, including the elapsed time between occurrences.

As described in the introduction to this chapter, the primary focus of our research questions is on the intensity of shift points about these two transition processes and what may trigger them. Therefore, we recorded the position of the last shift point (i.e., the starting point) and the position of the next shift point (i.e., the stopping point) for every shift point that occurred about mission analysis and tactical strategy. The starting and stopping points were scaled over the total number of communicated messages for the team during the episode in which they occurred.

In the dataset, teams could have multiple records, where one record represented each iteration of the topic shift points that occurred. Other data points within each record were the same across all records. For example, if a team had 3 shift points for mission analysis, then it would have 3 records, one including the starting and stopping points for each shift point. If the same team had 2 shift points for operational strategy, then 2 would be repeated in both records for the variable representing the number of operational strategy events. A more thorough description and demonstration of how the dataset was constructed can be found in this chapter's appendix.

As described in Chap. 3, cost and time have been scaled so that higher is better. That is, each team's cost performance was scaled by dividing the cost associated with the schedule they submitted by the optimal cost performance (i.e., 2438 in the first performance episode and 2430 in the second performance episode). The time performance was scaled by subtracting the time a team required to complete their schedule from the time recorded for the slowest team plus 1. In the first performance episode, the slowest team took 162 minutes, so a team that took 63 minutes was represented as 163–63 = 100. In the second performance episode, the slowest team took 160 minutes and therefore the team's time performance were determined by subtracting their recorded times from 161.

Analyses

Proportional Hazards Analysis

We use the methods outlined in Allison (1995) to conduct proportional hazards analysis in SAS 9.4. Where no shift point was found for a topic, the teams were right-censored in the analysis. We report the results only for models that are significant. This decision resulted in reporting results for tactical strategy, mission analysis, operational strategy, and action. The goal specification model was not significant, which was not surprising given that few goal specification shift points were identified in the dataset. This pattern of results was the same for both unrestricted and time pressured teams.

The dependent variables in our models are the transition processes and action. Independent variables include cost and time performance (note, time performance is only included for unrestricted teams because time pressured teams were given a fixed deadline, which resulted in very little variability in time performance), media, and performance episode. Additionally, we include two variables about each transition or action process that is not the dependent variable: (1) the Shift Point variables represent *when* the shift point occurs, as described in a previous section, and (2) the DDVs, labeled First Shift, indicate *if* the first shift point of the dependent variable occurred before the first shift point of the independent variable. For example, in the model where tactical strategy is the dependent variable, the Mission Analysis Shift First indicates whether or not the first tactical strategy shift point occurred before (Mission Analysis Shift First = 0) or after (Mission Analysis Shift First = 1) the first mission analysis shift point. Together these independent variables facilitate teasing out the importance of *when* and/or *if* one topic shift point occurred relative to when the shift point of the dependent variable occurred, respectively.

In our results tables, we report hazard ratios. For continuous variables (e.g., shift point), the hazard ratios describe the instantaneous rate of an event (i.e., dependent variable shift point) occurring. In other words, these ratios represent the probability that the event will occur at a specific moment in time, given that the event has not yet occurred. Hazard ratios greater than 1 indicate that events are more likely to occur, whereas ratios less than 1 indicate that events are less likely to occur. To aid in interpreting results, we employ the transformation 100∗(ratio-1) to ascertain the estimated percent change in the risk of an event occurring for a one-unit increase in the independent variable. For example, in the model with tactical strategy as the dependent variable, the hazard ratio associated with the operational strategy shift point is 0.396 and significant. We transform this value to 100∗(0.396–1) = −60.4%, which can be interpreted as the likelihood of the first tactical strategy shift point occurring decreases by 60.4% with every unit increase in when the operational strategy shift point occurs. In other words, if you want to increase the probability of having a tactical strategy shift point, the team needs to shift from operational strategy as early as possible.

For binary variables (e.g., shift first) the hazard ratio indicates the risk of the event occurring where the variable equals 1, compared to the condition where the variable equals 0. The same transformation can be used for easier interpretation, but the meaning provides the percent likelihood of an event happening versus not happening. Continuing the earlier example, the operational strategy shift first hazard ratio is also significant at 2.204 in the model with tactical strategy as the dependent variable. Transforming this value suggests that teams experiencing an operational strategy shift point are 120% more likely to also experience a tactical strategy shift point.

Andersen-Gill Intensity Analysis

Andersen-Gill intensity analysis, as devised by Andersen and Gill (1982), is used to evaluate recurrent tactical strategy and mission analysis shift points. The analyses were conducted in SAS 9.4.

The dependent variables in our models are mission analysis and tactical strategy. Independent variables include cost and time performance (note, time performance is only included for unrestricted teams), media, and performance episode. Additionally, shift point variables for all transition and action process, except for the one that is the dependent variable, are included as independent variables; theses variables represent when the first shift point occurs and are operationalized in the same way as they were for the proportional hazards analysis. The last set of independent variables included, labeled Events, are the numbers of shift points experienced by teams for each transition process (except for the dependent variable). The event variables help to determine if the number of shift points on topics other than the dependent variable affect the intensity of topic shifts teams experience with the dependent variable.

In our results tables, we report hazard ratios. These ratios represent the intensity functions associated with the dependent variables. Intensity, in our research, represents the repetition of topic shift points, where higher intensity indicates that teams experience more repetitions. Interpretation of these ratios is analogous to the hazard ratios that result from the proportional hazards analyses. For ratios less (greater) than 1, the dependent variable is less (more) intense if the independent variable has occurred.

Findings for Team Communication Polyphony

The descriptive statistics and correlations for teams working with unrestricted time or working with time pressure, including the performance episodes, media, shift point variables, and performance are shown in Table 6.1 below. Data for the unrestricted and time pressured teams can be found in the bottom or top of the matrix triangle, respectively. Note that time performance is only relevant for teams with unrestricted time.

Examining the Polyphony of Initial Conversations

Interrelationships Among Initial Topic Shift Points

Interrelationships Among Initial Topic Shift Points Results

The results of our examination of the interrelationships among initial topic shift points for both unrestricted and time pressured teams are presented in Table 6.2. We can answer our research questions by examining the patterns of results in this table.

Table 6.1 Polyphony descriptive statistics and correlations

	Unrestricted *M(SD)*	Time pressure *M(SD)*	1	2	3	4	5	6	7	8	9
1. Cost (scaled)	0.89 (0.09)	.80 (0.15)	-	NA	0.64[a]	0.09	-0.15	-0.07	-0.08	0.14	-0.16[c]
2. Time (scaled)	99.88 (25.70)	NA	0.40[a]	-	NA	NA	NA	NA	NA	NA	NA
3. Episode	0.48 (0.50)	.50 (0.50)	0.86[a]	0.49[a]	-	0	-0.23[a]	-0.10	-0.27[a]	-0.01	-0.13
4. Medium	0.47 (0.50)	.48 (0.50)	0.07	0.28[a]	-0.05	-	-0.10	0.17[c]	-0.09	0.06	0.06
5. Action	0.15 (0.15)	.20 (0.16)	0.03	0.05	-0.02	0.14	-	0.01	0.16[c]	-0.07	0.10
6. Goal specification shift point	0.01 (0.09)	.01 (0.03)	-0.17[b]	0.003	-0.04	0.03	-0.04	-	0.13	-0.05	-0.10
7. Mission analysis shift point	0.19 (0.21)	.20 (0.22)	-0.14	-0.20[a]	-0.21[a]	-0.16[c]	0.10	0.05	-	-0.07	0.13
8.Operational strategy shift point	0.15 (0.27)	.06 (0.18)	-0.04	-0.09	-0.04	-0.25[a]	-0.11	-0.07	0.15[c]	-	0.06
9. Tactical strategy shift point	0.22 (0.23)	.18 (0.19)	-0.34[a]	-0.27[a]	-0.27[a]	-0.21[a]	-0.02	0.01	0.17[b]	0.07	-

[a] $p < 0.01$; [b] $p < 0.05$; [c] $p < 0.10$

$N_{unrestricted}$ =135 teams. Correlations are shown in the bottom triangle of the matrix.

$N_{time\ pressured}$ =116 teams. Correlations are shown in the upper shaded triangle of the matrix.

Episode: first performance episode = 0; second performance episode = 1

Medium: distributed teams = 0, in person teams = 1

Table 6.2 Proportional hazards analysis results

	Unrestricted teams				Time pressured teams			
	Model 1: Action process	Model 2: Mission analysis	Model 3: Operational strategy	Model 4: Tactical strategy	Model 5: Action process	Model 6: Mission analysis	Model 7: Operational strategy	Model 8: Tactical strategy
Cost performance	0.736	2.182	0.139	725.493^{a}	3.743	0.288	0^{b}	13.680^{b}
Time performance	1.002	0.998	0.992	1.003	-	-	-	-
Episode	0.688	1.708	2.749	0.918	1.331	2.015^{c}	31.283^{b}	0.729
Medium	0.627^{b}	1.558	0.627	1.533^{c}	1.202	2.046^{a}	0.466	1.024
Action process shift point	-	0.089^{b}	35.532^{c}	0.078^{b}	-	0.272	1.009	0.048^{a}
Goal specification shift point	1.895	0.042^{b}	327.492^{b}	4.168	51.845	0.085	0	2006.106^{c}
Mission analysis shift point	0.582	-	0.165+	0.648	0.447	-	2.663	1.561
Operational strategy shift point	1.932^{c}	0.716	-	0.396^{b}	0.763	2.657^{c}	-	0.399
Tactical strategy shift point	1.007	1.266	0.640	-	0.641	0.974	0.310	-
Action first (DDV)	-	0.712	3.026	0.870	-	0.568	1.396	0.534^{c}
Mission analysis shift first (DDV)	2.304^{a}	-	0.971	0.968	3.324^{a}	-	0.337	2.306^{b}
Operational strategy shift first (DDV)	1.183	1.465	-	2.204^{b}	2.876^{a}	1.300	-	0.855
Tactical strategy shift first (DDV)	9.977^{a}	1.595^{c}	0.917	-	6.904^{a}	0.671	1.911	-
Model fit statistic	69.28(DF=11)^{a}	19.651(DF=11)^{b}	20.550 (DF=11)^{b}	39.372(DF=11)^{a}	98.260(DF=10)^{a}	18.678(DF=10)^{b}	21.428 (DF=10)^{b}	25.613(DF=10)^{a}

$^{a}p < 0.01$; $^{b}p < 0.05$; $^{c}p < 0.10$

$N_{unrestricted}$ =135 teams; 70 teams from the first performance episode and 65 teams from the second performance episode

$N_{time\ pressured}$ =116 teams; 58 teams from the first performance episode and 58 teams from the second performance episode

Episode: first performance episode = 0; second performance episode = 1

Medium: distributed teams = 0, in person teams = 1

Dynamic Dummy Variables (DDV): first shift point did *not* occur before dependent variable shift point occurred = 0; first shift point did occur = 1;

Notes: Chi-Square statistic to test Model Fit estimated using the Likelihood Ratio test.

The first research question focuses on the existence of interrelationships among initial topic shift points. As can be seen in the table, every model except the operational strategy model for time pressured teams includes significant results for topic shift points and/or the first shift DDVs. To answer the second research question about differences between the unrestricted and time pressured teams, we note the different patterns of significant relationships for the models with the same dependent variables. The third and fourth research questions are aimed at identifying any differences attributable to media and relationships with performance, respectively. Media is significant for three of the eight dependent variables tested, cost performance is significant for the tactical strategy models, and time performance is not significant for any models. Together, these findings suggest that topic shift points do form polyphonic patterns of interrelationships, and these polyphonic patterns differ based on time availability and communication medium used, and relate to team performance. Specific results are reported in the next sections followed by a summary of the patterns observed.

Unrestricted Teams

For unrestricted teams, when the shift points representing the start of action occur (Model 1) may be influenced in multiple ways. First, the action start shift point is related to what medium is used for communication; that is, the likelihood of initiating action was lower for in person teams than for distributed teams ($h = 0.627$, $p = 0.03$), suggesting that distributed teams initiated action sooner than in person teams. Second, when teams experienced a first operational strategy shift point influences when they will initiate action ($h = 1.932$, $p = 0.09$). Applying the transformation $100*(1.932–1) = 93.2$ we see that for each one-unit increase in when the first operational strategy shift point occurs the likelihood of initiating action goes up by 93.2%. Third, teams are also more likely to initiate action if they have experienced their first tactical strategy shift point ($h = 9.977$, $p < 0.0001$) or mission analysis shift point ($h = 2.304$, $p = 0.001$). Thus, the timing associated with initiating action is related to the media used, *when* teams experience their first operational strategy shift point, and *if* they have had meaningful conversations about tactical strategy and mission analysis.

The communication about mission analysis analyzed in Model 2, may be motivated by when teams discuss other topics. Experiencing a first mission analysis shift point is 59.5 times more likely if a first tactical strategy shift point has occurred ($h = 1.595$, $p = 0.09$). For each one-unit increase in when teams complete their first discussion of goal specification and in when they initiate action, the likelihood of their first mission analysis shift point goes down by 95.8% ($h = 0.042$, $p = 0.02$) and 91.1 % ($h = 0.089$, $p = 0.02$), respectively. In other words, delaying conversations about goal specifications or waiting to initiate action reduces the likelihood of teams having meaningful conversations about mission analysis. As such, experiencing a first mission analysis shift point is affected *if* teams have a tactical strategy shift point and *when* teams start action or experience a shift from goal specification.

The investigation into when the first operational strategy shift point (Model 3) occurs indicates that the timing is affected by when other shift points occur.

Specifically, the sooner teams have their first goal specification shift point ($h = 327.492$, $p = 0.007$) or start action ($h = 35.532$, $p = 0.06$), the more likely they are to quickly follow with their first operational strategy shift point. Conversely, the likelihood of having the first operational strategy shift point occur decreases by 83.5% for each one-unit increase in the timing of the mission analysis shift point ($h = 0.165$, $p = 0.06$). In sum, the polyphonic patterns associated with *when* teams experience their first operational strategy shift point suggest that earlier conversations about goal specification and initiating action encourage conversations about operational strategy; simultaneously, teams may want to engage in an early conversation about mission analysis to increase the likelihood of having a first meaningful conversation about operational strategy.

Finally, the results from examining what may relate to the occurrence of the first tactical strategy shift point for unrestricted teams (Model 4) show a positive relationship between cost performance and the first tactical strategy shift point ($h = 725.493$, $p = 0.01$), such that for each one-unit increase in cost performance the likelihood of the first tactical strategy shift point occurring goes up by 72,449.3%. Put another way, teams with high cost performance are more likely to experience an earlier first tactical strategy shift point than teams with low cost performance. We found that in person teams were 53 times more likely to experience a first tactical strategy shift point than distributed teams ($h = 1.533$, $p = 0.08$), suggesting that in person teams have an earlier first tactical strategy shift point than distributed teams. Interestingly, both if and when teams have a first operational strategy shift point matters for when they experience a first tactical strategy shift point. A first tactical strategy shift point was 120 times more likely for teams that have already experienced a first operational strategy shift point than for those that have not ($h = 2.204$, $p = 0.04$). As well, the likelihood of a first tactical strategy shift point decreases by 60.4% for each unit increase in the timing of the first operational strategy shift point ($h = 0.396$, $p = 0.04$). In other words, given that the first operational strategy shift point has happened, teams are likely to experience a long delay before their first tactical strategy shift point occurs. Similarly, once action has been initiated, teams are likely to experience a long delay before their first meaningful conversation about tactical strategy ($h = 0.078$, $p = 0.04$). Stated another way, the more time that passes before an operation shift point occurs or action is initiated, the less likely teams are to discuss tactical strategy. Thus, communication about tactical strategy may be prompted by *if* and *when* the team discusses its operational strategy and *when* teams initiate action. Moreover, in person teams may engage in conversations about tactical strategy sooner than distributed teams and the earlier shift point may positively impact cost performance.

Time Pressured Teams

For time pressured teams, the investigation of shift points capturing action start (Model 5), shows that, for the most part, if other shift points occurred, then action start was prompted to occur. Specifically, the likelihood of initiating action is 590.4 times more likely if the tactical strategy shift point has occurred ($h = 6.904$,

$p < 0.0001$), 232.4 time more likely if the mission analysis shift point has occurred ($h = 3.324$, $p < 0.0001$), and 187.6 times more likely if the operational strategy shift point has occurred ($h = 2.876$, $p = 0.008$). Thus, for time pressure teams the action start shift point is highly connected to their discussions about other topics.

For teams under time pressure, the mission analysis shift point was found to be prompted by media type and performance episode as well as when operational strategy shift point occurs (Model 6). The likelihood of experiencing the mission analysis shift point was 104.6 times greater for in person teams than for distributed teams ($h = 2.046$, $p = 0.005$). Teams were more likely to complete their first meaningful conversation about mission analysis earlier in the second performance episode than in the first ($h = 2.015$, $p = 0.06$). Further, the risk of a mission analysis shift point increases for each unit increase in the timing of the operational strategy shift point ($h = 2.657$, $p = 0.1$). As such, teams are more likely to have their first discussion about mission analysis soon after completing their first conversation about operational strategy.

The timing of the first operational strategy shift point analyzed (see Model 7) is related to performance such that higher performing teams are not likely to experience an operational strategy shift point ($h = 0$, $p = 0.02$). The timing of the first operational strategy shift point may also differ across performance episodes. Specifically, experiencing a first operational strategy shift point was found to be 3028.3 times greater for teams in the second performance episode than for teams in the first performance episode ($h = 31.283$, $p = 0.01$). This finding suggests that teams discussed operational strategy much sooner in the second performance episode than they did in the first.

Finally, the analysis of tactical strategy shift points for time pressure teams (Model 8) follows a similar pattern of results as for unrestricted teams. First, a positive relationship between the first tactical strategy shift point and cost performance ($h = 13.680$, $p = 0.02$) indicates that for each one-unit increase in cost performance the likelihood of the first tactical strategy shift point goes up by 1268%. As such, teams with better cost performance are more likely to engage in a meaningful first conversation about tactical strategy sooner than teams with lower cost performance. The timing, however, may not be as early as unrestricted teams given the large difference in the hazard ratios (unrestricted: $h = 725.493$; time pressured: $h = 13.680$). The first tactical strategy shift point is more likely to occur if teams have experienced their first mission analysis shift point ($h = 2.306$, $p = 0.02$); but it may be less likely to occur if the team initiates action ($h = 0.534$, $p = 0.09$). The timing of first topic shift points may also influence when the teams have their first conversational shift from tactical strategy. Specifically, the likelihood of the first tactical strategy shift point occurring increases for each unit increase in the timing of the first goal specification shift point ($h = 2006.106$, $p = 0.09$), but decreases for each unit increase in when action is initiated ($h = 0.048$, $p = 0.005$). Together, these findings highlight that the communication about tactical strategy for teams under time pressure may be influenced by *if* the team experiences a mission analysis shift point or initiates action, and *when* the team discusses its goal specification or action, which in turn impacts cost performance.

Overall Patterns Observed in Interrelationships Among Topic Shift Points

Our results suggest that completing the first meaningful conversation about tactical strategy may influence cost performance, regardless of the time available to complete the task. As such, teams may benefit from understanding what may prompt tactical strategy shift points. When working in an unrestricted condition, a first conversation about tactical strategy is more likely to occur earlier when meeting in person, so distributed teams may need to make a concerted effort to discuss their tactical strategy. Conversations about operational strategy and initiating action may need to be done earlier in the team's time together in order to provide adequate time for the team to discuss tactical strategy. Alternatively, time pressured teams may want to discuss mission analysis and goal specification early because both topics are likely to delay the occurrence of tactical strategy. These teams may also benefit from jumping right into action as getting some experience with the task may help them construct a better tactical strategy.

A second noteworthy pattern is that time pressured teams complete their first discussions about mission analysis and operational strategy more quickly during the second performance episode than they do in the first performance episode. These topics focus on what the team has been asked to accomplish and who is going to do the requisite work, respectively. Since the task in the second performance episode is only slightly different than the task in the first, this result is not surprising. Teams will have come to the second performance episode with task-related experience that informs their conversations about mission analysis and facilitates the assignment of work.

Our third point of interest is the complex relationship among the transition processes and the initiation of action. For both unrestricted and time pressured teams, either if or when mission analysis, operational strategy, and tactical strategy were discussed, spurred the team to action. One plausible conclusion from this polyphonic pattern is that teams may need to talk about all facets of their work and try different options through action before fully understanding their mission and settling on their final strategies. This notion is examined later in this chapter.

Examining the Polyphony of Recurring Conversations

To answer RQ6.5, we needed to determine if teams experienced multiple shift points (i.e., events) over time. Our dataset includes records for 135 unrestricted teams; these teams were communicating either through distributed media or in person and may be from either the first or second performance episode. Table 6.3 reports the numbers of unrestricted teams that experienced multiple events for tactical strategy and mission analysis. The data indicates that of the 135 examinations, all teams had at least one shift point recorded for both topics. Of these teams with at least 1 shift point found, 93 teams had at least 2 tactical strategy shift points and of these 35 teams had at least 3 tactical strategy shift points and so on. Interestingly, one team experienced 8 mission analysis events.

Table 6.3 Intensity (numbers of events)

	Unrestricted teams		Time pressured teams	
Event number	Mission analysis	Tactical strategy	Mission analysis	Tactical strategy
1	135	135	116	116
2	91	93	83	87
3	34	35	29	39
4	11	11	10	13
5	3	3	5	4
6	3	-	1	3
7	1	-	-	1
8	1	-	-	-

Frequencies include the total number for all distributed and In personTeams across both performance episodes

Similarly, we can see from the table that time pressured teams also experienced multiple shift points. In particular, of the 116 teams examined, all teams have at least 1 event for both mission analysis and tactical strategy. Of the 116 teams having a tactical strategy event, 87 had at least 2, and so on, with one team having 7 tactical strategy events. These results suggest, in answer to our research question, that teams do, indeed, repetitively iterate through topic shift points for both tactical strategy and mission analysis.

Intensity of Mission Analysis and Tactical Strategy Conversations

Intensity of Mission Analysis and Tactical Strategy Conversations Results

Our results examining the intensity of mission analysis and tactical strategy shift points for both unrestricted and time pressured teams are presented in Table 6.4. By examining the result patterns in the table, we can answer RQ6.6 through RQ6.9. RQ6.6 focuses on identifying what conditions or other topics may trigger repeatedly discussing mission analysis or tactical strategy. As can be seen in the table, all tested models have significant independent variables that influence topic intensity (i.e., how repetitively teams iterate through conversations about the topics of interest). The specific triggers will be discussed in subsequent sections. To answer RQ6.7, we compare the triggers between unrestricted and time pressured teams to discover some similarities (e.g., mission analysis events beget tactical strategy events and vice versa) and differences (e.g., performance episode is only significant for time pressured teams). The differences across communication media are the focus of RQ6.8. The results show that media is only significant for time pressured teams. Finally, RQ6.9 is aimed at ascertaining any relationships between polyphonic patterns of intensity and team performance. Time performance is significantly related to the intensity of both mission analysis and tactical strategy for unrestricted teams. Together these findings suggest that polyphonic patterns of intensity occur, have similarities and differences across time availability, differ across communication media only for time pressured teams, and impact time

Table 6.4 Andersen-Gill analysis results

	Unrestricted teams		Time pressured teams	
	Model 1: Mission analysis	Model 2: Tactical strategy	Model 3: Mission analysis	Model 4: Tactical strategy
Cost performance	6.823	0.118	1.276	1.593
Time performance	0.993[c]	0.989[a]	-	-
Episode	0.670	1.626	0.444[a]	0.618[c]
Medium	1.007	1.161	2.564[a]	1.859[a]
Action process shift point	1.274	0.151[a]	0.324[c]	0.298[b]
Goal specification shift point	0.440	21.637	0.028	0.005
Mission analysis shift point	-	0.595	-	1.270
Operational strategy shift point	1.292	0.901	1.676	0.402
Tactical strategy shift point	1.265	-	4.250[a]	-
Goal specification events	1.135	0.250	1.884	1.728
Mission analysis events	-	1.246[a]	-	1.245[a]
Operational strategy events	0.959	0.818	0.937	1.441[c]
Tactical strategy events	1.327[a]	-	1.451[a]	-
Model fit statistic	24.795[b]	34.795[a]	56.213[a]	44.565[a]

[a] $p < 0.01$; [b] $p < 0.05$; [c] $p < 0.10$

$N_{unrestricted}$ = 279 mission analysis events, 277 tactical strategy events

$N_{time\ pressured}$ = 244 mission analysis events, 263 tactical strategy events

Episode: first performance episode = 0; second performance episode = 1

Medium: distributed teams = 0, in person teams = 1

Notes: Chi-Square Score statistic to test Model Fit estimated using the Likelihood Ratio test.

performance in unrestricted teams. Specific results are reported in the next sections followed by a summary of the patterns observed.

Unrestricted Teams

In the examination of mission analysis intensity (see Model 1 in Table 6.4), the results indicate that teams earning better time performance had lower mission analysis intensity ($h = 0.993$, $p = 0.08$). As well, the intensity of mission analysis is increased by the number of tactical strategy events ($h = 1.327$, $p = 0.002$), such that for each one-unit increase in the number of mission analysis shift points that a team experienced, tactical strategy intensity increased by 32.7%.

The tactical strategy intensity results (see Model 2) indicate that, similar to mission analysis intensity, teams earning better time performance had fewer meaningful conversations about tactical strategy ($h = 0.989$, $p = 0.004$) such that for each one-unit increase in time performance the likelihood of engaging in iterative conversations about tactical strategy decreased by 1.1%. Teams that initiated action were less likely to repetitively discuss tactical strategy ($h = 0.151$, $p = 0.008$). Yet, the intensity of tactical strategy was increased by the number of mission analysis events ($h = 1.246$, $p = 0.0006$), such that for each one-unit increase in the number of mission analysis shift points, tactical strategy intensity increased by 24.6%.

Time Pressured Teams

The mission analysis (see Model 3) and tactical strategy (see Model 4) intensity exhibited by time pressured teams have similar relationships across the communication media being used and the performance episode in which the team is collaborating. Intensity associated with mission analysis and tactical strategy is 156.4% ($h = 2.564$, $p < 0.0001$) and 85.9% ($h = 1.859$, $p = 0.004$) higher, respectively, for in person teams when compared to distributed teams. The conversations during the second performance episode were less intense than the conversations during the first performance episode (mission analysis: $h = 0.444$, $p = 0.002$; tactical strategy: $h = 0.618$, $p = 0.06$).

The intensity of mission analysis is also influenced by three additional constructs. The later a team's tactical strategy first shift point occured ($h = 4.250$, $p = 0.003$) and the more tactical strategy events that transpired ($h = 1.451$, $p = 0.0002$), the more likely teams were to have iterative discussions about mission analysis (i.e., mission analysis intensity is higher). The intensity of mission analysis is decreased by when the action shift point occurred ($h = 0.324$, $p = 0.08$), such that for every one-unit increase in when teams first shifted to action, the likelihood of mission analysis intensity decreased by 67.6%.

The intensity of tactical strategy is decreased by when an action shift point occurred ($h = 0.298$, $p = 0.05$), which suggests that starting action earlier could generate greater tactical strategy intensity. Tactical strategy intensity increased as the numbers of meaningful conversations about mission analysis ($h = 1.245$, $p = 0.009$) and operational strategy ($h = 1.441$, $p = 0.06$) increase.

Overall Patterns Observed in Intensity of Mission Analysis and Tactical Strategy Shift Points

When considering whether intensity of discussions matter, we find that for unrestricted teams, the intensity of tactical strategy and mission analysis matter in terms of team efficiency. That is, if they engaged in fewer repetitive discussions about mission analysis and tactical strategy then they had better time performance. In other words, by streamlining their conversations, they required less time to complete their tasks. Therefore, teams may want to consider strategies to foster such streamlined communication such as initiating action later to reduce the intensity of tactical strategy. Even though the time performance of time pressured teams was not assessed, these teams must be cognizant of elapsing time because of their time constraints. Therefore, they may also benefit from reducing the intensity of tactical strategy and delaying the initiation of action.

Regardless of the time available, teams need to be aware that conversations about mission analysis precipitate conversations about tactical strategy, and vice versa. This entanglement between mission analysis and tactical strategy is further examined in Chap. 7 when we focus on how teams are discussing these two transition processes.

Two interesting patterns can be seen across the results from the proportional hazards and the Andersen-Gill intensity analyses. First, the proportional hazards analyses results suggest that if and when the first tactical strategy meaningful conversation occurs has an impact on cost performance, regardless of time availability. For the Andersen-Gill intensity analyses, we observe that time performance may be impacted by the intensity of those conversations in unrestricted teams. Thus, both when teams begin talking about their transition processes and how frequently they have those discussions must both be considered to achieve good performance. Second, experienced time pressured teams demonstrated more succinct discussion behaviors in the second performance episodes by completing their first meaningful conversations about mission analysis and operational strategy quickly and iterating fewer times about tactical strategy and mission analysis. Examining how they converse about these topics in Chap. 7 may offer guidance that will be beneficial for all teams.

Theoretical Implications and Future Research

Our results offer three insights that confirm and extend Marks and colleagues (2001) recurring phase model of team processes. First, our results identify when teams may be best served discussing certain topics, which answers Marks and colleagues (2001) call to identify transition-action timing within performance episodes. Such evidence "is critical to understanding not only what but when team processes become critical to goal accomplishment" (p. 361). Our results extend this notion by also uncovering the polyphonic patterns that demonstrate the interrelationships among topics. So, in addition to identifying when meaningful conversations about transition processes and the initiation of action occur, our results also depict the manner in which certain topics prompt other topics. Second, we confirm Marks and colleagues (2001: 369) conceptualization that teams "engage in multitask processing" by demonstrating the interrelationships among the topic shift points. In particular, our results focused on the initiation of action, strongly suggest that teams are engaging in multi-topic conversations about transition processes as they are testing ideas through action. Finally, we further extend our observation, first published in Kennedy and McComb (2014), that Marks et al. (2001) strategy formulation transition process may be better operationalized as two distinct transition processes. As can be seen in our results, teams devise tactical and operational strategies through different communication patterns. Future researchers may benefit from examining strategy formulation from these two perspectives.

Two additional theoretical frameworks may be applicable to the research conducted to examine team communication polyphony. First, bounded rationality (Simon 1972) may help us explain the different conversational approaches employed by unrestricted and time pressured teams. Unrestricted teams engage in more complete conversations as they work their way through what needs to be done and how it should be accomplished before assigning work to members. Without limits on the time they have to accomplish their work, these teams have no constraints on their ability to follow a systematic process to discover, elaborate, and evaluate what they are doing. In other words, they engage in a rational approach as they collaborate on their task. Alternatively, time pressured teams may function using a bounded rational approach. These teams include action early in their time together as they iterate through trial-and-error approaches to determine the best tactical strategy. The time constraints placed on these teams appear to limit how systematic they can be and, at times, may compel them to employ satisficing strategies to move forward expeditiously.

The second theoretical framework that may be informed by our research findings is Fiori and colleagues' (2010) conceptualization of macrocogition in teams. Through their framework, they suggest that team members build upon their internalized knowledge as they collaborate. Our topic shift points, and the interrelationships among them, may be useful in understanding this process. Specifically, Fiori and colleagues (2010) identify shared mental models as one component of internalized team knowledge. As the content of the team members' mental models converges over time, these individually-held structures become more similar, or shared, thereby building their internalized knowledge. Once the team members have shared mental models, they shift from actively focusing on the mental model content to subconsciously employing that knowledge as they engage in their taskwork until a time when the shared mental model may need updating (McComb 2007). Kennedy and McComb (2010) embedded communication into the mental model convergence process to demonstrate how communication might be useful as a window into the cognitive processes of team members. The topic shift points, operationalized as meaningful conversations followed by extended periods of conversations about other topics, aligns with McComb's (2007) conceptualization of the mental model convergence process. Therefore, the interrelationships among, and intensity of, topic shift point may offer insights into the macrocognitive functioning of teams.

Given the importance of shared mental models to team performance, research explicitly linking the mental model convergence process and team communication may be warranted. Other opportunities to advance our examination of polyphony include efforts to more holistically examine team processes, in general, and team communication, in particular. By employing event history analyses, we have evidence regarding when, if, and how often teams communicate about various topics. This type of analytical approach could be expanded to include other components of team communication such as externalized knowledge development (e.g., physical artifacts) or focus on other team processes such as behavioral coordination mechanisms.

Methodological Contributions and Future Research

From a methodological perspective, our use of two components of event history analysis, namely proportional hazards and Andersen-Gill intensity models, allowed us to present a holistic view of team communication polyphony that included not only if and when topics are discussed, but also how often they are discussed. Event history analysis is particularly salient for nuanced examinations of longitudinal team communication because, in comparison to standard regression analysis, it can handle time-varying covariates and censored data with limited information loss. Although not novel in the examination of longitudinal team communication (see McComb et al. 2010), the application is limited. Our efforts herein extend the work of McComb and colleagues (2010) by incorporating an examination of iterations about specific topics. The attention to iterative cycles and what prompts them may generate new ideas about communication management and interventions. Thus, event history analysis offers researchers a host of tools that can facilitate learning about longitudinal team communication and its impact on performance.

Much of the longitudinal research about longitudinal team communication has been conducted using ethnographies or other types of observation studies that collect rich data over time, but then aggregate it to represent themes. In many cases, this aggregation transforms longitudinal data into cross-sectional data. Researchers may benefit from thinking more longitudinally and temporally as they consider how they want to conduct their analyses. As highlighted in this chapter, event history analysis includes a myriad of analytical approaches (Allison 1984). Our efforts demonstrate the power of combining approaches to tell a more complete story. Future researchers may look to the plethora of event history analysis options available to conduct their own nuanced examination of team communication and other longitudinal team processes.

Managerial Minutes

Engage in iterative, meaningful conversations.
Teams need to have a meaningful conversation about how they are going to complete their task. Engaging in these conversations early helps teams create a foundation for their assigned work. Revisiting these conversations periodically is also necessary to ensure that the team continues to move in the right direction.

Watch for team communication that spirals out of control.
Understanding the mission and creating a strategy to accomplish it are interwoven. Teams need to acknowledge this interconnection, but not let themselves spiral out of control. Too many iterations may lengthen the time required to finish the task.

Appendix: How to Conduct Event History Analysis

In Chap. 6, we examined the polyphony of team communication using event history analysis. This approach provided insights about the temporal interdependencies among when topics are discussed and the intensity with which topics are revisited. To demonstrate how we employed event history analysis, in general, and the proportional hazards and the Andersen-Gill intensity models, in particular, we evaluate the example dataset presented in the Chap. 3 Appendix. The pattern recognition and measurement required to complete these analyses were conducted in Matlab. Statistical analyses were conducted in SAS 9.4.

Data Preparation

The first step in data preparation for event history analysis is to define the event in which you are interested. For our purposes, the event was the end of a sustained conversation about a particular topic or the initiation of a conversation about a topic. In Chap. 6, the pattern of communication that led up to an event was determined by subject matter experts. These experts used an iterative process to ascertain the appropriate number of messages to be exchanged and the number of subsequent messages about other topics that would best represent a completed conversation. This process was informed by an evaluation of the raw messages, insights about pattern capture for different settings, and an understanding of the team literature. For the example dataset in this appendix, we set the pattern recognition rule for topics 10, 20, and 30 to be a minimum of 3 sequential messages on a topic followed by at least 2 messages about other topics, which parallels the pattern used in Chap. 6 for the transition processes. We treat topic 40 as a topic similar to our action topic in Chap. 6, where our interest is in when it was initiated.

The next step is to determine the event scores for each event of interest. To obtain the event scores, we evaluated each communication string using our pattern rules in Matlab. For topics 10, 20, and 30, the event was set at the last message about the topic code of interest exchanged before the topic was changed for an extended period of time. For topic 40, the event occurred when the first message coded 40 was exchanged. When a pattern did not occur in the communication string, the event was scored a 0. Because all communication strings were of different lengths, all event scores were scaled by dividing the raw score by the total number of messages exchanged in the communication string.

The communication string, with corresponding position numbers, for Team 1 in our example dataset is:

Position:	Topic:	Position:	Topic:
1	10	11	10
2	10	12	10
3	10	13	10
4	20	14	40
5	20	15	40
6	20	16	40
7	20	17	40
8	30	18	40
9	30	19	40
10	30	20	40

For this team, the pattern recognition rule for the topic code 10 is found at position 3 (i.e., 3 messages coded 10 are exchanged and followed by at least 2 messages about other topics). The event score is then calculated as the position divided by the total number of messages in the communication string, or 10Pattern = 3/20 = 0.15. For this team, the event scores for the other topics would be 20Pattern = 0.35, 30Pattern = 0.50, and 40Pattern = 0.70. The compiled dataset used for the proportional hazard analysis is shown in Table 6.5.

The Andersen-Gill intensity model provides greater insight about the repeated pattern occurrences (i.e., intensity) for topics of interest. In our example dataset, we focus on the intensity of topic 10. We obtain the event scores needed for analysis by extending the pattern recognition rule beyond just the first occurrence. Using this approach, we capture every occurrence of the pattern that exists in the communication string. Each event score requires a starting and stopping point. The first starting point is 0 and the last stopping point is 1. Starting and stopping points between the 0 and 1 represent the scaled positions on the communication string when the events occur, where scaling is done by dividing the position where the event occurred by the total number of messages in the communication string.

For Team 1 in the example dataset the event scores for topic code 10 show that the pattern is found twice, once as noted above where the sequential messages end at position 3 (scaled position = 0.15), and again where the sequential messages end at position 13 (scaled position = 0.65). Therefore, the starting and stopping points for Team 1's topic 10 would be: [0, 0.15], [0.15, 0.65], and [0.65, 1]. Each event is entered into the dataset as its own record. In other words, every event is recorded in its own row.

Although we elected to focus on the intensity of topic 10 as the dependent variable, we also determined that it may be related to the first occurrence of a pattern for topics 20 and 30 and how many times the pattern occurs for those topics. For Team 1, topic 20 and topic 30 patterns are found once each and the event scores are 0.35 and 0.5, respectively. Note that these values are repeated for every iteration (i.e., row) of a topic 10 pattern. A full account of the dataset is shown in Table 6.6.

Table 6.5 Example teams proportional hazards model dataset

Team	10Pattern	20Pattern	30Pattern	40Pattern
1	0.15	0.35	0.5	0.70
2	0.45	0.23	0.64	0.82
3	0.50	0.21	0.63	0.79
4	0.14	0.45	0.59	0.64
5	0.50	0.17	0.71	0.54
6	0.32	0.50	0.14	0.36
7	0.20	0	0.35	0.40
8	0.32	0.45	0.18	0.05
9	0.36	0.68	0.50	0.18
10	0.14	0.27	0.41	0.73

Table 6.6 Example teams Andersen-Gill intensity model analysis dataset

Team	10Pattern Start	10Pattern Stop	Pattern End	10Event Number	20Pattern (when)	30Pattern (when)	40Pattern (when)	20Event (number of)	30Event (number of)
1	0	0.15	0	1	0.35	0.5	0.7	1	1
1	0.15	0.65	0	2	0.35	0.5	0.7	1	1
1	0.65	1	1	3	0.35	0.5	0.7	1	1
2	0	0.45	0	1	0.22727	0.63636	0.8818	2	1
2	0.45	1	1	2	0.22727	0.63636	0.8818	2	1
3	0	0.5	0	1	0.20833	0.625	0.79167	1	1
3	0.5	0.75	0	2	0.20833	0.625	0.79167	1	1
3	0.75	1	1	3	0.20833	0.625	0.79167	1	1
4	0	0.13636	0	1	0.45455	0.59091	0.63636	1	1
4	0.13636	0.77273	0	2	0.45455	0.59091	0.63636	1	1
4	0.77273	1	1	3	0.45455	0.59091	0.63636	1	1
5	0	0.5	0	1	0.16667	0.70833	0.54167	1	1
5	0.5	1	1	2	0.16667	0.70833	0.54167	1	1
6	0	0.31818	0	1	0.5	0.13636	0.36364	1	2
6	0.31818	0.77273	0	2	0.5	0.13636	0.36364	1	2
6	0.77273	1	1	3	0.5	0.13636	0.36364	1	2
7	0	0.2	0	1	1	0.35	0.4	0	1
7	0.2	0.7	0	2	1	0.35	0.4	0	1
7	0.7	1	1	3	1	0.35	0.4	0	1
8	0	0.31818	0	1	0.45455	0.18182	0.04545	1	2
8	0.31818	0.77273	0	2	0.45455	0.18182	0.04545	1	2
8	0.77273	1	1	3	0.45455	0.18182	0.04545	1	2
9	0	0.36364	0	1	0.68182	0.5	0.18182	1	1
9	0.36364	1	1	2	0.68182	0.5	0.18182	1	1
10	0	0.013636	0	1	0.27273	0.40909	0.72727	2	1
10	0.13636	0.54545	0	2	0.27273	0.40909	0.72727	2	1
10	0.54545	0.90909	0	3	0.27273	0.40909	0.72727	2	1
10	0.90909	1	1	4	0.27273	0.40909	0.72727	2	1

Data Usage

Proportional Hazards Model

In our demonstration of the proportional hazards model from event history analysis, we assess the relationship between when the pattern for a topic of interest occurs in relationship to when patterns of other topics occur. The descriptive statistics and intercorrelations among our variables for this analysis are shown in Table 6.7.

The proportional hazards model relates time-varying covariates to the dependent variable. The resulting analysis provides information about the hazard ratio, which represents the likelihood of an event occurrence given that another event has occurred while holding other variables constant. The reciprocal of the hazard ratio is the expected time until the event occurs. The equation for event history analysis is:

$$h(t) = h_0(t) \times \exp(\beta_1 x_1 + \beta_2 x_2 + \ldots + \beta_p x_p)$$

where, t represents the survival time and the set $(x_1, x_2, \ldots, x_p)$ represents the covariates.

To aid in interpreting results, we offer the following rubric.

- Employ the transformation 100∗(ratio-1) to get the estimated percent change in the likelihood of the dependent variable occurring for a one-unit increase in the independent variable.
- If the hazard ratio is >1, events are more likely to occur. If the hazard ratio is <1, events are less likely to occur.
- If the independent variable is continuous, the dependent variable will be increasingly more/less likely to occur as the independent variable increases. For example, the older you get, the more likely you are to die.
- If the independent variable is binary, the dependent variable is more/less likely to occur if the independent variable =1 than if the independent variable = 0. For example, someone who has been incarcerated is more likely to go back to jail than someone who has never been to jail.

The results of the analysis using our example dataset are shown on the right side of Table 6.7. Model 1, where the outcome variable is the 10Pattern, is not significant ($\chi^2 = 3.003$, $p = 0.39$). Model 2, where the outcome variable is the 20Pattern, is significant ($\chi^2 = 7.831$, $p = 0.05$). In Model 2 we see that the hazard ratio of the 40Pattern is significant ($h = 271.590$, $p = 0.05$). This result suggests that the timing of the 20Pattern depends on the timing of the 40Pattern. Specifically, because the 40Pattern is a time-varying covariate, we may say that for every one-unit increase in when the 40Pattern occurs, the likelihood of the 20Pattern increases 270.590 times (100∗(271.590–1) = 270.590). That is, the 20Pattern is much more likely to occur once the team experiences a 40Pattern.

Model 3 is marginally significant ($\chi^2 = 7.548$, $p = 0.06$) and shows that when the 30Pattern occurs is linked to when the 10Pattern ($h = 0.001$, $p = 0.07$) and 40Pattern ($h = 0.014$, $p = 0.07$) occur. Interestingly, the results suggest that the 30Pattern is less likely to occur when the 10Pattern and 40Pattern occur because the hazard of occurrence goes down for every one-unit increase in these variables. This result may suggest that timing of the 30Pattern is later in the communication string if the 10Pattern or 40Pattern occur.

Finally, Model 4 is found to be marginally significant ($\chi^2 = 6.378$, $p = 0.09$). In this case, the 40Pattern is linked to the 30Pattern ($h = 0.006$, $p = 0.08$) where the hazard of a topic 40 message being exchanged goes down for every one-unit increase in when the 30Pattern occurs.

Table 6.7 Example teams descriptive statistics, correlations, and regression results for proportional hazards model analysis

	Descriptive statistics		Correlation results			Proportional hazards model regression analysis			
Variables	*M*	*SD*	1	2	3	Model 1 10Pattern	Model 2 20Pattern	Model 3 30Pattern	Model 4 40Pattern
1. 10Pattern	0.31	0.15	-			-	318.380	0.001[b]	3.971
2. 20Pattern	0.33	0.20	-0.10	-		3.506	-	0.520	66.069
3. 30Pattern	0.46	0.19	0.36	-0.29	-	0.019	1.915	-	0.006[b]
4. 40Pattern	0.52	0.26	0.01	-0.48	0.64[a]	6.024	271.590[a]	0.014[b]	-
Chi-Sq (df)						3.003 (3)	7.831 (3)[a]	7.548 (3)[b]	6.378 (3)[b]

[a] $p < 0.05$; [b] $p < 0.10$

Andersen-Gill Intensity Model

To demonstrate our use of the Andersen-Gill intensity model, we focus on topic 10 and how the other events may affect how frequently the topic 10 patterns recur. The descriptive statistics and correlations for the Andersen-Gill analysis are shown in Table 6.8. The 10Start and 10Stop variables represent the starting and stopping points for each recurrence of a topic 10 pattern. The 20Pattern, 30Pattern, and 40Pattern variables represent the scaled point in the communication string when the pattern of interest occurred. Finally, the 20Event and 30Event variables represent how many times the patterns for topics 20 and 30 were found in the communication strings, respectively.

To conduct the analysis, we added two additional variables to the dataset: (1) PatternEnd, which is a binary variable indicating whether or not it is the end of 10Pattern recognition (i.e., 1 when it ends, 0 otherwise) and (2) 10EventNumber, which incrementally counts the number of rows (i.e., number of start/stop patterns identified in a communication string) for each team. For Team 10, this 10EventNumber = 1 in the first row, 2 in the second row, and 3 in the third row.

The Andersen-Gill intensity model equation is:

$$h_z(t)\mathrm{d}t = E\{\mathrm{d}N(t) \mid F_{t-}\} = h_0(t)\exp[\beta' x(t)]\ \mathrm{d}t$$

where $N(t)$ is the number of events by a subject over the interval $[0, t]$.

The interpretation of the hazard ratios follows the same guidelines as provided above for the proportional hazards model. In addition to regression results for the Andersen-Gill intensity model (see right columns in Table 6.8), the output also includes a summary of the number of teams experiencing multiple events. As can be seen in Table 6.9, of 10 teams in the example dataset that could have experienced one event, all 10 did have at least one topic 10 pattern in their communication strings. Of those 10 teams experiencing one event, 7 teams experienced a second topic 10 pattern in their communication string and 3 teams or 30% were censored (i.e., they did not have a second topic 10 pattern in their communication strings).

Table 6.8 Example teams descriptive statistics and correlations for Andersen-Gill intensity model analysis

	Descriptive statistics		Correlations							Andersen-Gill intensity model regression results	
Variable	*M*	*SD*	1	2	3	4	5	6		Outcome	Model 1 10Pattern
1. 10Start	0.32	0.31	-							20Pattern	0.001[b]
2. 10Stop	0.67	0.33	0.87[a]	-						30Pattern	0[a]
3. 20Pattern	0.43	0.24	-0.06	-0.06	-					40Pattern	0.250
4. 30Pattern	0.45	0.18	-0.09	0.03	-0.44[b]	-				20Event	1.001
5. 40Pattern	0.53	0.26	0.03	-0.01	-0.56[a]	0.67[a]	-			30Event	0[b]
6. 20Event	1.11	0.57	0.04	0.00	-0.74[a]	0.19	0.47[b]	-			
7. 30Event	1.21	0.42	0.07	0.04	0.09	-0.83[a]	-0.68[a]	-0.10		Chi-Sq(df)	19.269 (3)[a]

[a]$p < 0.01$; [b]$p < 0.05$; [c]$p < 0.10$

Table 6.9 Summary of the number of events and censored values for example teams

Event number	Number of teams that could experience event	Number of teams that experienced event	Number of teams censored	Percentage of teams censored
1	10	10	0	0%
2	10	7	3	30%
3	7	2	5	71.43%
4	1	0	1	100%
Total	28	19	9	32.14%

The results in Table 6.8 show that the model for the intensity of the 10Pattern is significant ($\chi^2 = 19.269$, $p < 0.01$). The number of repetitive topic 10 patterns is significantly impacted by the number of topic 20 ($h = 0.001$, $p = 0.04$) and topic 30 ($h = 0.000$, $p < 0.01$) patterns, as well as the total number of topic 30 events ($h = 0.443$, $p = 0.44$). In all cases, topic 10 patterns are less likely to occur because the significant hazard ratios are all less than 1.

Chapter 7
Sequence of Team Communication

Team leaders and members often have varying levels of aptitude for teamwork. Some may, for example, readily engage in team conversations or take on specific roles that help facilitate team momentum. Others may struggle with understanding their purpose on the team or engaging effectively in team conversations. Team leaders may want to encourage certain behaviors that foster an effective team climate while working individually with struggling team members to ensure they do not undermine team activities. The communication sequences examined in this chapter may prove useful in identifying different types of communication behaviors that may be most advantageous to teams. Specifically, we investigate how teams converse about our topics of interest by classifying both the topic (e.g., tactical strategy, mission analysis) and purpose (e.g., clarify, suggest, inform) of each exchanged message, examining the sequential order in which these messages are transmitted, and determining any differences in sequence attributable to the conditions under which the team is working or their performance on the assigned task. Capturing how team communication unfolds at this level of detail will highlight for team leaders the types of messages or ordering of messages that may facilitate or impede team progress, thereby offering guidance about any interventions that might be warranted.

Introduction to Team Communication Sequencing

Through our examination of communication rhythm, timing, pacing, and polyphony in the previous chapters, we have offered guidance about *what* teams discuss and *when* they discuss it. In this chapter, we dig deeper into *how* teams discuss these topics through an examination of the sequences that comprise the teams' conversations. Sequence, as defined by Bartunek and Woodman (2015: 168) in their conceptualization of organizational change, represents the "temporal ordering of events … within a particular stage or phase." In our team-level investigation of sequence,

S. McComb, D. Kennedy, *Computational Methods to Examine Team Communication*, Computational Social Sciences,
https://doi.org/10.1007/978-3-030-36159-4_7

those events are the communicated messages exchanged among team members and the stages or phases are the quartiles across the two performance episodes.

To incorporate how teams discuss the topics, we apply a secondary layer of coding. These codes, described in Chap. 3, provide information about a team member's purpose for exchanging a particular message (i.e., to act, agree, apply, clarify, inform, suggest, or summarize). The result of this approach is that each message has a topic-purpose code (e.g., *tactical strategy-suggest, mission analysis-summarize*). Our results, therefore, uncover the manner in which sequential patterns across (topic) and within (purpose) the teams' conversations unfold over time.

The search for sequences in team interactions has been a long-time interest of team researchers and practitioners. Indeed, seminal work about group interaction structure considers how to frame the phases (e.g., Bales and Strodbeck, 1951, McGrath and Altman 1966), activities (e.g., Gladstein 1984; Hackman and Morris 1975) and sequential structures in group interaction (e.g., Poole 1983). These framings assert that temporal patterns of interaction exist. To garner insights beyond acknowledging what structures exist, however, requires a better understanding about why these structures may promote team success (Meyers and Seibold 2012). Our more granular investigation of team communication may help.

Researchers acknowledge that a "cookbook" does not exist describing how to study sequences within data about groups and teams (Hewes et al., 2012). As described in Chap. 2, many different approaches are feasible. In order to conduct our analyses, we selected hierarchical clustering to identify sequential structures in the communication strings and dendrograms to visually represent the topic-purpose clusters. Hierarchical clustering has been effectively used to discover sequential structures in the communication that occurred during team meetings (Lehmann-Willenbrock et al. 2016; Olson et al. 1994) and training exercises (Gorman et al. 2016). We conducted a preliminary assessment of the communication exchanged by the unrestricted in person teams in our sample (McComb et al. 2014). The data were coded with an initial content-based coding scheme, rather than our current process-based coding scheme. Nevertheless, the results demonstrate the applicability of this methodological approach for our purposes by suggesting that sequential structures exist in the communication strings and sequential patterns differ over time.

Our results throughout this book indicate that communication topics differed across quartiles and conditions (Chapters 4, 5 and 6), and that the order and timing of topic discussions may be related to team performance (Chapters 4 and 6). As described in those chapters, the literature also supports these differences. Herein, we build upon our findings by identifying differences in the topic-purpose sequential patterns across quartiles, that may be a function of the conditions under which the teams are collaborating (i.e., distributed versus in person, unrestricted versus time pressured), and/or between high and/or low performing teams.

Research Questions

The research questions associated with sequence are:

RQ7.1. Do teams have sequential patterns in their communication over time?
RQ7.2. Does the sequence of communication differ depending on the time teams have available (i.e., unrestricted time versus time pressured)?
RQ7.3. Does the sequence of communication differ depending on the communication media available to the team (i.e., distributed versus in person)?
RQ7.4. Do sequential patterns in team communication relate to cost performance?

Methodology

Methodological Approaches: Hierarchical Clustering Analysis and Dendrograms

Hierarchical clustering analysis was used to identify sequential structures within our communication strings. We followed the procedures outlined by Olson et al. (1994), which includes three steps. First, chi-squared tests of independence were conducted to determine if our observed data is due to chance or contains sequential structures. These results provide an answer for RQ7.1. Second, clusters were established by using the distance between two objects (i.e., messages with topic-purpose codes) to identify natural groupings in the data, which are the hierarchical clusters. Third, dendrograms were constructed to visually represent the topic-purpose clusters. A dendrogram is a type of diagram that illustrates hierarchical relationships among elements, and can be used to garner insights about how embedded certain topic-purpose codes were in the discussions in relation to other topic-purpose codes. The results are used to answer RQ7.2, RQ7.3, and RQ7.4 because they provide more nuanced guidance about how teams converse over time. Moreover, they augment what we have uncovered about the rhythm, timing, pacing, and polyphony of team communication in previous chapters.

Data Specification

The data used for analyses were communication strings containing topic-purpose codes for all exchanged messages. In other words, each exchanged message in our communication strings was comprised of two pieces: (1) the topic of the exchange, which could be action, goal specification, mission analysis, operational strategy, or tactical strategy and (2) the purpose, which could be action, agree, apply, clarify, inform, suggest, and summarize. Two frequency tables, one for each performance episode, were constructed that included data from all teams. These frequency tables contained the total number of every possible topic-purpose transition, where transitions occur from one topic-purpose exchange that happens at time t to a topic-purpose exchange that happens at time $t + 1$. For example, in our dataset for the first performance episode, across all media and time conditions, the teams in our sample completed a total of 732 transitions from *tactical strategy-suggest* to *tactical strategy-agree*. The totals for every possible transition, by performance episode, are recorded in the frequency tables.

Analyses

Chi-Squared Tests of Independence

Chi-squared tests of independence, conducted in Microsoft Excel, were used to ascertain if our data contained sequential structures. This analysis started with an initial assessment of all teams by performance episode to make general decisions about what topic-purpose transitions would be retained for the analyses required to answer our research questions. We followed the work of Olson et al. (1994) in considering the removal of transitions with low or zero expected values. These removals help us to avoid violating the assumptions associated with chi-squared tests and maintain parsimony. Additionally, we considered our previous findings from earlier chapters and theoretical interest in Marks et al.'s (2001) transition and action processes. The combination of these considerations resulted in five manipulations to the dataset. First, all goal specification transitions were removed because they rarely occurred (approximately 0.25% of all exchanges were about goal specification during both performance episodes). Second, operational strategy represented approximately 1% and 2% of all exchanges during the first and second performance episode, respectively. Yet, our results, particularly in Chap. 6, suggest that operational strategy plays a role in team conversations. Therefore, we decided to keep operational strategy in our analyses, but only at the topic level (i.e., all messages were collapsed into one topic-level category for operational strategy). Third, we decided to hone our focus on transition processes, which led to collapsing all action messages into one topic-level category.

Our fourth manipulation resulted in omitting two purposes from the mission analysis and tactical strategy topics given their low frequency of occurrence. Specifically, all transitions that included action (0.3% and 0.12% of exchanges during the first and second performance episodes, respectively) and apply (0.04% of exchanges during both performance episodes) were removed from the dataset.

Our fifth and final manipulation was to divide our dataset into 64 subsets based on the results of our earlier analyses. Specifically, we compare across the two media types (distributed and in person) and two time availability conditions (unrestricted and time pressured), which results in four types of teams. Additionally, we divided the teams by their cost performance, using a median split, which resulted in two performance groupings per type of team. Finally, we broke the communication strings into quartiles across the two performance episodes (eight quartiles).

Given these decisions we prepared datasets with 12 categories (i.e., action and operational strategy are topic-level categories; mission analysis and tactical strategy each have five topic-purpose codes). When testing if the structure of our frequency data differed from expected frequency, however, additional categories had to be removed because they had low or zero frequencies. We elected to make these removals on a case-by-case basis because they differed across team types and quartiles, which is not unexpected given our earlier results suggesting differences across these dimensions. Even with these omissions, we retained 96.9–100% of exchanges for our analyses.

Hierarchical Clustering Analysis

Hierarchical clustering analysis and dendrogram construction was completed in Matlab. The built-in functionality in the Statistics and Machine Learning Toolbox was used to conduct the hierarchical cluster analysis. The cluster connections were determined using Euclidean distance. This measure tells us about the similarity or dissimilarity between topic-purpose transitions. In our analyses the distances were exponentially varied and therefore we used the square root of the distance as the height in order to facilitate better visual comparisons across clusters.

Next, we graphically displayed the hierarchical clusters using dendrograms. The dendrogram essentially links topic-purpose transitions with similar patterns in a tree structure. An example dendrogram is given in Fig. 7.1. All dendrograms can be viewed online in Mueller et al. (2019). Here, we describe how to interpret the dendrograms. The dendrogram is constructed such that topic-purpose codes are listed along the horizontal axis. The order changes based on the hierarchical structure in the data. The height of the hierarchical tree depicted in the dendrogram is based on the distance between clusters, the value of which is represented along the vertical axis. As such, those topic-purpose clusters at the bottom of the dendrogram have low distance and those that are separated by taller lines are very distant.

To systematically discuss the insights from the dendrogram the next step is to determine where to cut the clusters and consider all objects below that cut distance as a single cluster. Determining the most appropriate cutoff distance is a qualitative, context dependent decision that can range along a continuum from aggregating all clusters into a single cluster to interpreting each individual cluster. After reviewing the dendrograms produced, we explored cutoffs of 2, 4, 6, and 8 to ascertain which cutoff value balanced our interests in discerning multiple clusters and facilitating interpretability. Our investigation resulted in a cutoff distance of 6, as it produced a distinguishable and meaningful number of clusters to discuss.

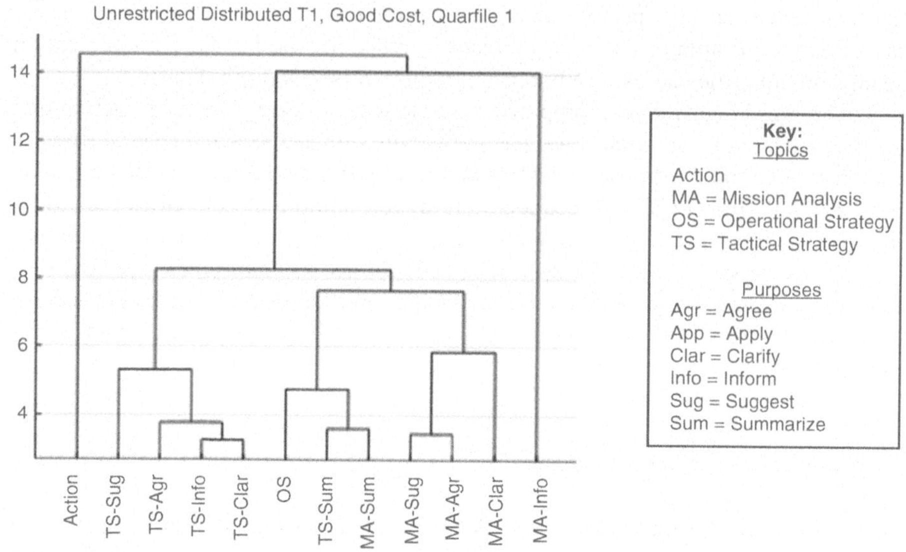

Fig. 7.1 Example dendrogram

Findings for Team Communication Sequencing

Determining If Sequential Patterns Exist

To determine if the data structures were different than random, chi-squared tests of independence were run on all of the 64 matrices constructed to represent each of the four quartiles, for teams with good or bad cost performance, working either distributed or in person, and having unrestricted time or being time pressured. The Chi-squared test statistics ranged from 486.78–3763.57; all tests were significant at $p < 0.0001$. Thus, in a preliminary answer to RQ1, we found that team communication does contain sequential patterns. The overall patterns observed in our results, which we describe in this section, provide a more refined answer to this research question.

Examining Sequences in Team Communication

Sequential Patterns in Team Communication

Sequential Patterns in Team Communication Results

Table 7.1 summarizes our findings from the 64 dendrograms representing each of the four quartiles, for teams with good or bad cost performance, working either distributed or in person, and having unrestricted time or being time pressured.

Table 7.1 Hierarchical clustering analysis summary

Condition	Q1	Q2	Q3	Q4	What good cost teams did differently than bad cost teams in this condition
Unrestricted, distributed, T1	G: 5 clusters B: 5 clusters	G: 4 clusters B: 5 clusters	G: 2 clusters B: 3 clusters	G: 2 clusters B: 3 clusters	TS early MA over time
Unrestricted, distributed, T2	G: 5 clusters B: 5 clusters	G: 4 clusters B: 4 clusters	G: 2 clusters B: 3 clusters	G: 2 clusters B: 2 clusters	TS/OS early OS over time
Unrestricted, in person, T1	G: 6 clusters B: 5 clusters	G: 4 clusters B: 3 clusters	G: 3 clusters B: 2 clusters	G: 3 clusters B: 2 clusters	MA early TS over time
Unrestricted, in person, T2	G: 4 clusters B: 4 clusters	G: 2 clusters B: 2 clusters	G: 2 clusters B: 2 clusters	G: 2 clusters B: 2 clusters	MA early TS over time
Time pressure, distributed, T1	G: 4 clusters B: 5 clusters	G: 3 clusters B: 3 clusters	G: 2 clusters B: 2 clusters	G: 2 clusters B: 2 clusters	TS early TS/MA over time
Time pressure, distributed, T2	G: 4 clusters B: 4 clusters	G: 2 clusters B: 3 clusters	G: 2 clusters B: 2 clusters	G: 2 clusters B: 2 clusters	TS early TS over time
Time pressure, in person, T1	G: 6 clusters B: 5 clusters	G: 5 clusters B: 4 clusters	G: 3 clusters B: 3 clusters	G: 2 clusters B: 3 clusters	MA/OS early TS over time
Time pressure, in person, T2	G: 4 clusters B: 5 clusters	G: 2 clusters B: 3 clusters	G: 2 clusters B: 2 clusters	G: 2 clusters B: 2 clusters	MA early TS over time

G = good cost teams; B = bad cost teams
MA = mission analysis; OS = operational strategy; TS = tactical strategy
Note: In all conditions across all quartiles, one cluster focuses on action; the remaining clusters focus on transition processes.

Specifically, the table provides information about how many clusters are observed for teams with good and bad cost performance across quartiles for all conditions. Additionally, a summary of what the good cost performance teams did differently than teams with bad cost performance is provided. For example, unrestricted distributed teams with good cost performance discuss their tactical strategy earlier than bad cost teams and discuss mission analysis across all quartiles. Only the dendrograms for the first performance episode are included in this chapter (see Fig. 7.2). All dendrograms are available online in Mueller et al. (2019).

In the assessment of clusters across all conditions we see three general patterns worth noting that provide the evidence needed to answer our research questions. First, across all conditions, the numbers of clusters per quartile becomes smaller, which indicates that the similarity among transition patterns increases over time. In other words, as teams progress, how and what they discuss becomes streamlined. Combining these results with the aforementioned chi-squared test results, our answer to the first research question is that sequential patterns exist in team communication and they differ over time.

Second, differences do exist in the sequential patterns of unrestricted and time pressured teams' conversations, but those differences may be dependent upon media condition. In distributed teams, those facing time pressure had fewer clusters in the first two quartiles than their unrestricted counterparts, suggesting that distributed teams were more focused in their conversations than in person teams. These results affirm that teams differ depending on time availability (RQ7.2) and communication media (RQ7.3).

Third, we observed two general trends in how many clusters good and bad cost teams have over time, one for each media condition. In distributed teams, those teams with good cost performance typically have fewer clusters than those with bad cost performance. Conversely, in person teams with good cost performance tend to have more clusters in a quartile than teams with bad cost performance. These patterns suggest that sequential patterns may differ between good and bad performing teams in response to our fourth research question. But those differences may be dependent upon the media through which they are conversing, providing further evidence that differences in sequential patterns exist across communication media (RQ7.3).

The above observations are all based on the numbers of clusters observed. Nevertheless, they provide evidence of differences across time, conditions, and performance in response to our research questions. More nuanced examinations of these differences that incorporate the topic-purpose transitions are provided in the following sections. In particular, we focus our observations on the sequential patterns used by good cost performance teams during the first performance episodes and contrast them with the patterns observed in the bad cost performance teams. We follow by offering summary observations about differences across performance episodes, between teams with different time availability, and between teams working with different communication media.

Overall Patterns of Results by Team Type

Unrestricted Distributed Teams

We begin by looking at unrestricted, distributed teams with good cost performance. The dendrograms for these teams during their first performance episode can be seen in Fig. 7.2a. The messages about transition processes for their first quartile falls into four clusters, these are:

1. *tactical strategy-inform*, *tactical strategy-clarify*, *tactical strategy-agree* and *tactical strategy-suggest*;
2. *tactical strategy-summarize*, *mission analysis-summarize* and *operational strategy*;
3. *mission analysis-suggest, mission analysis-agree* and *mission analysis-clarify*;
4. *mission analysis-inform.*

Each cluster is examined, including substantiation from transcript data, to understand its role in the orientation of teams to their task.

First, to discuss tactical strategy, teams engage in similar sequential patterns where they provide information about working on the task, clarify that they understand the tactical strategy being considered, suggest possible approaches to complete the task, and agree on what they want to do to get the task done. For example, in a transcript of an unrestricted distributed team with good cost performance we found that the team discussed tactical strategy by providing information "there are 10 workers, if each person work 4 hours a day, there are 40 hours," clarifying the strategy "wait, so have we decided on how many hours per employee…?" suggesting what to do "let's make sure all the high wages people get all the hours, penalty will be less for those lower wages workers," and agreeing on the strategy "yep."

The good cost teams discuss mission analysis in two distinct clusters. The first cluster focuses specifically on exchanging mission analysis information, which makes sense since all members had unique rules regarding the scheduling task that

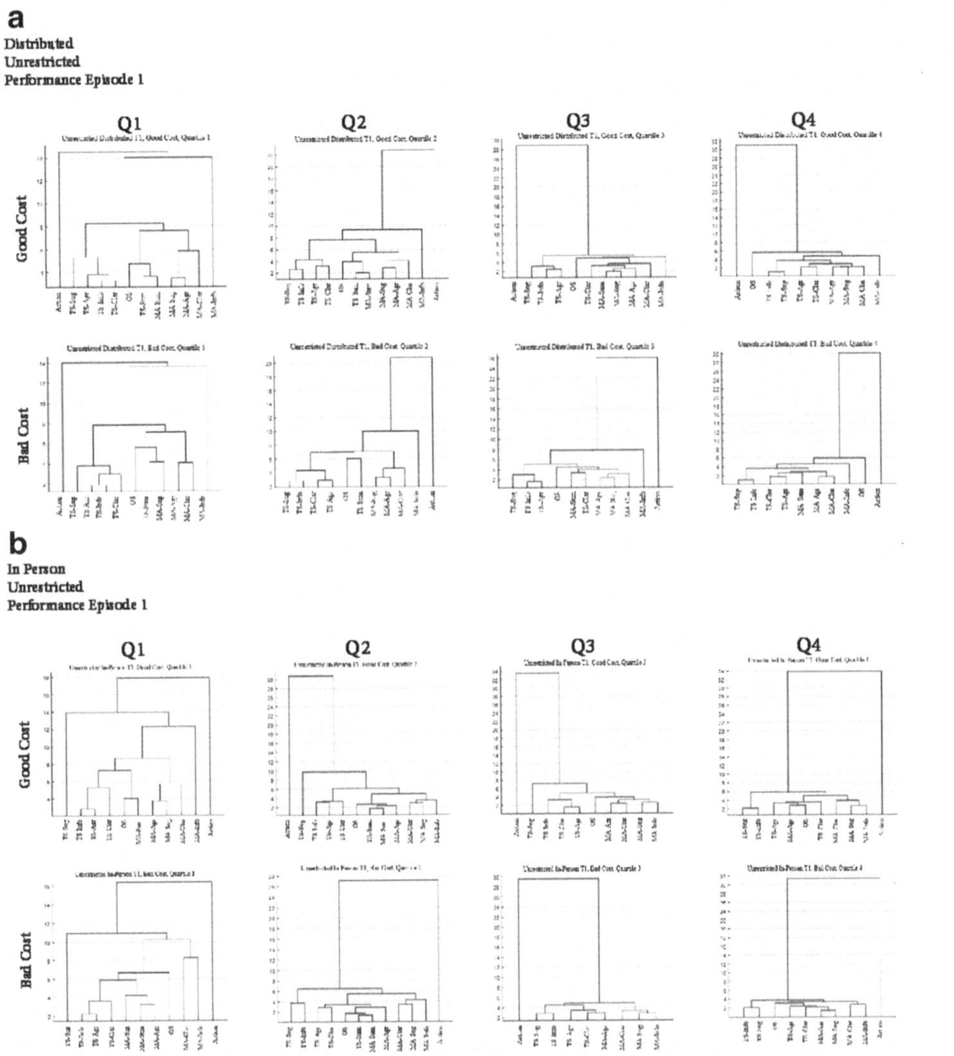

Fig. 7.2 Dendrograms for the first performance episode by condition. (**a**) Dendrograms for unrestricted distributed teams in performance episode 1, (**b**) Dendrograms for unrestricted in person teams in performance episode 1, (**c**) Dendrograms for time pressured distributed teams in performance episode 1 and (**d**) Dendrograms for time pressured in person teams in performance episode 1

invoked interdependence among them. For example, one team focused on reporting out the information "okay, mine says day 1: 3 ppl per shift, 2: 3 ppl/shift, 3: 5 ppl / shift, 4: 4ppl/shift, 5: 6 ppl/shift 6: 7 ppl/shift, 7: 4 ppl per shift" another member added "yes, we have the same info here" and the team continued to compare task parameters. In the other cluster focused on mission analysis, teams ensured that all members understand the parameters guiding what they have been asked to do by

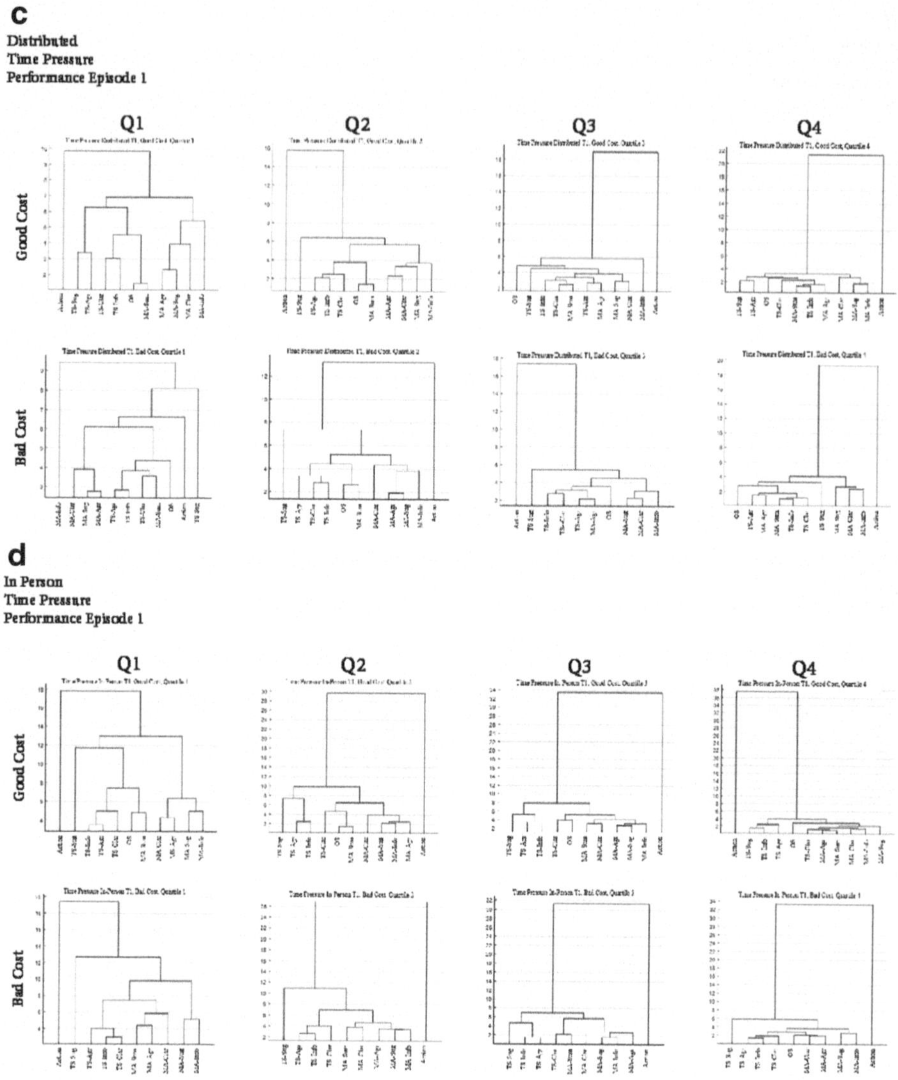

Fig. 7.2 (continued)

exchanging messages about mission analysis to clarify, suggest and agree. For example, "okay, so is your penalty 1.8 times the extra hours worked times the wage?" with the response of "yep."

In the final cluster, the teams similarly transitioned among summaries of tactical strategy and mission analysis as well as operational strategies, which may have helped the team toward closure about these two topics and prepared them to act. For instance, one team brought together the tactical strategy and mission analysis by synthesizing their parameters and summarizing "it's 4 to 10 hours each day" and making the connection to strategy "it's only 8 to 8pm, 12 hours a day...," "yep, so at some point we switch the workers."

During the first quartile, unrestricted distributed teams with bad cost performance have some similarities with those achieving good cost. The most obvious similarity is the independent cluster focused on exchanging information about their mission. Two noteworthy differences, however, may help explain the performance difference. First, in discussions about tactical strategy, the sub-cluster with inform also has agree in the bad cost performance teams, whereas in the good cost performance teams the other exchange type is clarify. This difference suggests that lower performing teams may be reporting an approach before realizing why they are taking it. For example, one team transcript contains exchanges about strategy information (e.g., "I am going high to low," "I am just jotting down things so i can see what it will cost us"), whereas some time passed before a member asked for a clarification by writing, "so what's your strategy here?" The response was a suggestion of "no low wages on day 6, we have to use up the hours for the least wages first," which prompted one member to seek clarification by asking, "where do u see that?" and then the team moved into a discussion of mission analysis. The team's conversation about mission analysis provided information and clarification through statements such as "we have to have three in each box all day" followed by "I don't have that sheet ...what does yours say?" and "nothing like that...where is it?" The ongoing confusion may have inhibited their ability to succeed.

The second difference between the good and bad cost teams has to do with summarizing mission analysis, which is missing from the teams with bad cost performance. The good cost performance teams have a cluster of *tactical strategy-summarize*, *mission analysis-summarize* and *operational strategy.* In the bad cost performance teams *mission analysis-summarize* is replaced with *mission analysis-suggest*, which begs a question about whether suggesting parameters is less helpful than summarizing them. For example, as one team member attempts to implement a strategy by summarizing "we have to use up the hours for the least wages first...are we all on the same page here?" another member countered with the mission analysis suggestion "it doesn't have to be the same people all day." And when a team member tried to establish an operational strategy by stating, "I will keep track of how many hours people are working...you fill in the boxes," subsequent messages suggest that other team members had lingering questions about the tactical strategy ("no wait, let's get the minimum hours for the high wages") and the mission analysis ("does every single employee need to work every day?"). In other words, the team hadn't effectively connected what they were doing to why they were doing it.

Beyond the first quartile, we continue to see similarities and differences in the sequential patterns exhibited by teams with good and bad cost performance. In the second quartile, good and bad cost teams have two similar transition clusters with a distance score less than 6:

1. *tactical strategy-inform*, *tactical strategy-agree*, *tactical strategy-clarify*, and *tactical strategy-suggest,*

2. *mission analysis-inform.*

These two clusters suggest that all teams are touching on the approaches they are taking while continuing to share mission analysis information. The patterns of differences observed between good and bad cost teams build on the differences identified in the first quartile. Good cost teams now have one cluster that includes: *tactical strategy-summarize, mission analysis-summarize, operational strategy, mission analysis-suggest, mission analysis-agree,* and *mission analysis-clarify.* Collapsing what was two cluster in the first quartile into one cluster, suggests that the good cost teams are referring to their strategy, the rules, and the distribution of work throughout the quartile as they engage in action. Bad cost teams continue to have two clusters:

1. *mission analysis-suggest, mission analysis-agree,* and *mission analysis-clarify,*
2. *tactical strategy-summarize and operational strategy.*

Moreover, they continue to lack *mission analysis-summarize* messages. These clusters, combined with the omission of *mission analysis-summarize* suggest that these teams may continue to be unclear about their mission and their strategy to achieve it.

In the third quartile, both good and bad cost teams have fewer clusters. Good cost teams have one cluster that includes all topic-purpose transition processes at the distance of 6. This grouping suggests that these teams are "checking in" to make sure their plans and distribution of work are sound. For example, the action message "Gail didn't work at all in day 7, she needs at least 4 hours" was followed by the mission clarification "is that the rule?" and another action message where a member indicated "I can fix that." For bad cost teams, two clusters are present. Notably, in this quartile *mission analysis-summarize* is added to the conversation, but a *mission analysis-inform* cluster remains. These foci on mission analysis may suggest that bad cost teams still did not have a clear understanding of their task parameters. Indeed, this issue shows up in the transcripts, as demonstrated by the team member whose response to "does every single employee need to work every day?" was "yes, that's what I said 100 times."

The fourth quartile also contains a small number of clusters. For good cost teams, they maintain one cluster with all topic-purpose transition processes. Interestingly, fewer topic-purpose codes are represented suggesting that these teams streamline the way they talk about their transition processes. In particular, they no longer need to summarize the tactical strategy or mission analysis. Bad cost teams have two clusters, one focused on tactical strategy and mission analysis exchanges, and one focused on operational strategy. The separation of operational strategy suggests that these teams may still have work to do. Indeed, we found one bad cost team still distributing the bulk of the work in the fourth quartile saying "I'll do [day] 6 and 7" while another member indicates "I'll do [day 3 and 4] … I'll do 5 also." In addition, the bad cost teams continue to have *mission analysis-summarize* messages, suggesting they are still putting the pieces together at the end.

In summary, unrestricted distributed teams generally have fewer clusters over time. However, bad cost teams maintained more clusters throughout the task, notably having *mission analysis-inform* sustained as a cluster into the third quartile, whereas good cost teams integrated it with other topics after the second quartile. As well, bad cost teams did not exchange messages about *mission analysis-summarize* until the third quartile and then continued these messages into the fourth quartile, whereas good cost teams stopped these message patterns in the third quartile. Finally, we pointed out a nuanced difference in the way tactical strategy was discussed in the first quartile, where good cost teams clarify and bad cost teams agree as they exchange information about their tactical strategy. This subtle difference may offer one opportunity to catch teams early and ensure that the members are connecting how they are going to approach their task and why they are using that approach (inform and clarify).

Unrestricted In Person Teams

Next we look at the conversations of unrestricted in person teams (see Fig. 7.2b). Considering those clusters with distance less than 6, the clusters about transition processes in the first quartile for good cost teams are:

1. *tactical strategy-suggest,*
2. *tactical strategy-inform, tactical strategy-agree, tactical strategy-clarify,*
3. *operational strategy* and *mission analysis-summarize,*
4. *mission analysis-agree, mission analysis-suggest, mission analysis-clarify,*
5. *mission analysis-inform.*

Unrestricted in person teams repeatedly return their attention to tactical strategy; for example suggesting "so do you want to take the most expensive workers and put them in" and then "we should split them" and the further suggestion of "okay we'll put one every morning and one every night." The team similarly checks the strategy by asking "what's an important day" with responses of "Saturday because we need 7 people" and "7 times 4 is 28 alright so Ian and Jen we want them to work the least so if we could put them on together every day for four hours and one day for six hours Saturday day six." These types of discussions highlight how good cost teams are suggesting strategies and then making sure that the strategies are understood (info, agree, clarify).

The cluster of *operational strategy* and *mission analysis-summarize* indicate that teams are able to take what they have learned about their task through the *mission analysis-inform* cluster (e.g., "they can only work ten hours a day" and "no less than four") and formulate how to move forward. For example, one team summarized the parameters governing what they were tasked with doing (e.g., "so between 4 and ten hours a day") and then articulated how work could be assigned (e.g., "I'm going to keep track of the hours").

For the bad cost teams in the first quartile, the clusters are:

1. *tactical strategy-suggest,*
2. *tactical strategy-inform, tactical strategy-agree, tactical strategy-clarify, mission analysis-suggest, mission analysis-summarize, mission analysis-agree,*
3. *operational strategy,*
4. *mission analysis-clarify* and *mission analysis-inform.*

These clusters are somewhat different than those observed for good cost teams. Although the teams have some similarities (e.g., focused attention on suggesting strategies), one difference is particularly notable. The cluster where the teams clarify and provide information about mission analysis suggests that teams may be struggling to gather and understand all of the parameters governing their task. For example, a transcript shows the team clarifying "fifty hours a week, is that a maximum?" and "where does yours say that" with information such as, "I think that's the only one that has that."

In the second quartile, good cost teams have three clusters focused on their transition processes:

1. *tactical strategy-suggest,*
2. *tactical strategy-inform*, *tactical strategy-agree*, and *tactical strategy-clarify,*
3. *operational strategy, tactical strategy-summarize, mission analysis-summarize, mission analysis-agree, mission analysis-clarify, mission analysis-suggest,* and *mission analysis inform.*

These separate clusters suggest that good cost teams are still refining their tactical strategies. For example, one team member suggested, "we're not going to have Ian and Jen work as much as Al Bob and Ceci." This suggestion was followed by information and agreement about their team's mission analysis, "they all have to work thirty hours," "right," and followed with another suggestion about tactical strategy "well you want to keep them to thirty until everyone has their thirty and then start filling in the cheapest people."

In the second and fourth quartiles, the bad and good cost teams, respectively, have a separate cluster that includes *tactical strategy-suggest* and *tactical strategy-inform*. The reasons for which the teams use these clusters, however, is very different. In the second quartile, the bad cost teams may be struggling with their strategy. For example, one team's discussion included, "we are going to give everyone their minimum first" (suggest), "so it's going to be like the penalty is the same… "(inform), "I don't think this is going to work out…" (suggest), "look how it looks and we're at G how many people do we need there per day" (inform). The good cost teams in the fourth quartile appear to be working on improving their final schedule with this cluster as they offer suggestions such as, "so this one we can subtract …" or provide information to identify opportunities to improve like, "cause the top lines are [here]." These attempts to improve their schedules were not their first. The good cost teams had two clusters in the third quartile (i.e., *tactical strategy-suggest* and all others). They used the *tactical strategy-suggest* cluster to check and refine their strategy between discussions about action and other topics, through statements such as "would we want to do that because we want to make it as cheap as we can," and "we can go back and fix things." Whereas bad cost teams, with only one cluster about transition processes, rarely broke up their action to assess what they were doing.

In summary, like the distributed teams, these unrestricted in person teams generally have fewer topic-purpose clusters over time. Unlike in the distributed condition, however, bad cost teams had *fewer* clusters over time than the good cost teams. The key differences for in person teams were the clusters around tactical strategy. Good cost teams continued to introduce messages about tactical strategy by both suggesting and providing information throughout. This approach appears to facilitate refining their tactical strategies over time, as well as gaining buy-in about the strategy from all members. This evolving focus on tactical strategy may be one opportunity to facilitate (1) early conversations where teams hone their tactical strategy and (2) ongoing dialog where teams maintain, update, and/or refine their approach to the task as work progresses.

Time Pressured Distributed Teams

Next we focus on the differences between good and bad cost teams in the time pressure distributed context (see Fig. 7.2c). For good cost teams, we find three clusters with distance less than 6:

1. *tactical strategy-suggest* and *tactical strategy-agree,*
2. *tactical strategy-clarify, tactical strategy-inform, operational strategy* and *mission analysis-summarize,*
3. *mission analysis-agree, mission analysis-suggest, mission analysis-clarify,* and *mission analysis-inform.*

Interestingly, these cluster patterns are similar to unrestricted distributed good cost teams with one notable difference. Whereas unrestricted teams have the luxury of focused conversations about mission analysis and tactical strategy that they then summarized to facilitate crafting their operational strategy, time pressured teams jump into making assignments before having a fully formed tactical strategy that can be summarized. In other words, they replace *tactical strategy –summarize* with *tactical strategy-clarify* and *tactical strategy-inform*. For example, one member asked for clarification relating to a proposed tactical strategy by stating, "the only thing w/u'r idea [teammate name] is im not sure how to go about this … what do u mean assign people and days?" and the teammate reiterated the operational strategy as "that you do monday I do tues …".

For bad cost teams in the time pressure distributed context, four clusters about transition processes are identified:

1. *mission analysis-inform,*
2. *mission analysis-clarify, mission analysis-suggest,* and *mission analysis-agree,*
3. *tactical strategy-agree, tactical strategy-inform, tactical strategy-clarify, mission analysis-summarize,* and *operational strategy,*
4. *tactical strategy-suggest.*

These patterns are surprisingly similar to the unrestricted in person good cost teams with the *tactical strategy-suggest* and *mission analysis-inform* as separate clusters. Yet, this approach was not effective under time pressure in a distributed setting. Focusing on sequences of suggesting strategies and reporting the task parameters may be a poor use of time. For example, in one team over half of the exchanges during the first quartile were either *tactical strategy-suggest* or *mission analysis-inform*. This squandering of the limited time available does not efficiently move them toward a common understanding of their mission or agreement about tactical strategy that can be put into action.

In subsequent quartiles, the topic-purpose clusters of good and bad cost teams are quite similar. But, differences do emerge when the actual conversations are taken into consideration. That is, in the second quartile teams with good and bad cost performance have two clusters: *tactical strategy-suggest* and all others. The difference between the two types of teams lies in how the teams converse within the cluster containing all topic-purpose exchanges except *tactical strategy-suggest*. In particular, the role of *mission analysis-clarify* may be a key difference. In the teams with good cost performance, *mission analysis-clarify* statements are used to hone in on what may or may not be feasible given the parameters of the task, such as "can someone come in a 8-10 then come back from 4-6?" Alternatively, the bad cost teams continued struggling to grasp the rules by stating "so what now…" or "huh?" or "does your talk about the penalty?"

In the third and fourth quartiles the teams with good and bad cost performance have one cluster focused on transition processes. While in the third quartile teams only have one cluster at a cutoff distance of 6, both good and bad performance teams have one cluster just below the cutoff that is comprised of only one topic-purpose code. For good cost teams that cluster is *mission analysis-inform*, whereas for bad cost teams the cluster is *tactical strategy-suggest*. This difference is pivotal. The good cost teams are using the information about their mission analysis to check their work. For example, one member pointed out in a message about action that "day 2, A, E, J, H don't work" and then went on to provide information about the mission in support of that statement by stating, "well I need everyone to work at least 4 hours per day." Alternatively, the bad cost teams may still be trying to finalize an approach this late in the game (e.g., "we should use Fay and Gail only 30 hours per week") or even giving up (e.g., "maybe we should just fill them randomly to just have something done"). These sequential patterns continue into the fourth quartile.

In summary, as with unrestricted teams, time pressured distributed teams had fewer clusters over time. As the teams begin their collaborations, one key difference is in their use of the time available. Good cost teams jumped into creating an operational strategy before fully constructing their tactical strategy. Alternatively, the bad cost teams spent overlong exchanging information about mission analysis and offering suggestions about tactical strategy. A second difference between the teams with good and bad cost performance is in the specific content of their messages. Good cost teams appear to use their communication to hone their understandings and refine their strategies. Bad cost teams, alternatively, seem to repeat the same conversations in attempts to clear up ongoing confusion. Facilitating conversations focused on quickly attaining a basic understanding of the mission, which then serves as a foundation for planning, may be an opportunity to help teams accomplish a critical first step.

Time Pressured In Person Teams

Finally, we focus on the differences between good and bad cost teams in the time pressured in person condition (see Fig. 7.2d). For good cost teams, we find five clusters with distances less than 6:

1. *tactical strategy-suggest,*
2. *tactical strategy-agree, tactical strategy-clarify, tactical strategy-inform,*
3. *mission analysis-summarize* and *operational strategy,*
4. *mission analysis-agree* and *mission analysis-clarify,*
5. *mission analysis-suggest* and *mission analysis-inform.*

These teams kick off their collaboration in a manner similar to unrestricted in person teams with one difference. Specifically, *mission analysis-suggest* clusters with *mission analysis-inform* in the time pressured teams, whereas *mission analysis-inform* is in its own cluster in unrestricted teams. This shift suggests that the time pressure may encourage teams to engage in richer discussions about the mission analysis versus a straight information exchange that gets sorted out via a different cluster. For example, one time pressured team discussed how penalties worked by reading information about how a penalty would be calculated if a worker was not scheduled for enough hours, suggesting "that's if it's a penalty that's if you schedule them too much right?", providing more information by saying, "no that's if you only schedule him for only 3 hours a day, you have to you schedule them for", ending with the suggestion "so you need 4 hours a day."

The bad cost teams differ from good cost teams in two ways. First, none of their clusters include operational strategy. In fact, operational strategy does not appear until the fourth quartile. These teams have visual cues because they are working face-to-face, which may offset the need to formally articulate who is doing what. Nevertheless, the lack of formality may hinder their abilities to perform. Second, *mission analysis-summarize* is clustered with *mission analysis-agree* and *mission analysis-clarify*. This clustering suggests that lower performing teams use the summaries to help ensure a common understanding about their mission (e.g., "um does it say anything about hours per day my people can't work less than 4 hours a day" (clarify), "mine can't work more than 10" (inform), "so in between 4 and 10" (summarize)). Alternatively, the better performing teams used *mission analysis-summarize* to help determine their operational strategy.

Team communication is similar across the second and third quartiles, with the good cost teams engaging in conversations about mission analysis and tactical strategy in different clusters, but integrating *tactical strategy-clarify* and *operational strategy* with *mission analysis-summarize* (second quartile) or a broader conversation about mission analysis (third quarter). This sequential pattern suggests that as the teams are conducting their action, they are also clarifying that the work they are doing aligns with their strategies and mission analysis. For example, one team summarized that an action can be undertaken by stating, "just as long as it is between 30 and 50" and subsequently questioned, "are you sure that's the most efficient?" For the lower performing teams, they retain a cluster focused on *tactical strategy-*

suggest during the second quartile and the transcript review suggests that these teams continued to struggle with how to approach the task (e.g., "Alright. Should we try to figure something out until [time runs out]?" "Why don't we have the highest people or the lowest people? To balance itself out").

In the final quartile, the higher performing time pressured in person teams are focused primarily on their action, with all transition processes clustered together at very low distance scores. This structure suggests that any conversations about transition processes were quick checks to ensure the team was working according to their plans and within their parameters. Alternatively, the bad cost teams continued to have a unique cluster focused on *tactical strategy-suggest*. The isolation of this topic-purpose code so late in performance episode may be a sign of struggle that hindered their ability to achieve a good outcome. For example, one team ran through a series of suggestions about how to approach the task at the beginning of the fourth quarter including "or would it be better to take", "... why don't we just do one", "just randomly toss them in and we'll see how close we come to making it work".

In summary, the time pressured in person teams with good cost performance are similar to their distributed counterparts in that they engage in rich conversations to quickly grasp what they have been tasked to do. The bad cost performance teams, however, seem unable to garner a clear understanding of their mission, nor can they devise a workable tactical strategy. Indeed, the sequential patterns suggest that these teams continue to consider alternate strategies into their fourth quartile. Simultaneously, they rely on visual inference to allocate work among the team members by not formally discussing their operational strategy until the final quartile, which may suggest a missed opportunity to be clear about their distribution of work.

Comparing Sequential Patterns across Conditions

Comparisons Across Performance Episodes

When examining similarities and differences across the two performance episodes, three overarching themes can be seen. First, teams readily get into action during the second performance episode, particularly after the first quartile. As can be seen in Table 7.1, the numbers of clusters during the first two quartiles are lower in the second performance episode than in the first. Also, in all but the unrestricted distributed teams, the good cost teams have only one cluster focused on transition processes during the last three quartiles. These findings suggest that teams are able to apply learning from the first performance episode during the first quartile of the second performance episode to figure out how they were going to complete the task and then quickly get to work for the remaining time.

Second, the overall numbers of topic-purpose codes found in the communication strings reduced from the first to the second performance episode. These reductions were most pronounced in the second, third, and fourth quartiles of all in person teams that achieved good cost performance, regardless of time availability. Additionally, the fourth quartiles of time pressured distributed teams (both good and bad cost performance teams) had one cluster about their transition processes that included six and eight topic-purpose codes, respectively. Interestingly, the good cost teams' clusters included no tactical strategy codes. Across all of the quartiles with fewer topic-purpose codes, we see no *mission analysis-summarize* or *tactical strategy-summarize* messages; and *operational strategy* is missing from the third quartile. This observation connotes that teams may engage in more strategic communication as they become more familiar with each other and the task required of them.

Finally, the sequential patterns established by good and bad cost performance teams in the first performance episode continue into the second performance episode. For example, in the unrestricted distributed condition, the good cost teams settled what they were going to do during the first quartile and then checked to be sure they were on track for the remaining quartiles. For example, a team asks "so what's our strategy this time?" And the answer is "let's get it done as fast as possible so we can go back and make it as low-cost as possible." With the response "k that sounds cool." Alternatively, the bad cost performance teams during the second performance episode were still exchanging mission analysis information into the second quartile and establishing an operational strategy into the third quartile. For example, a bad cost performance team continues to ask "how should we do this?" with multiple responses such as "do you want to work each day together?" and "I think it's better to share results at the end." We did not provide any performance feedback to the teams after the first performance episode, which may help explain why the bad communication behaviors continued into the second. This theme indicates that a caveat to the aforementioned point about learning accomplished during the first performance episode may be necessary. Specifically, learning occurs over time, but not all teams will be learning better collaborative practices.

Comparisons of Distributed Versus In Person Teams

Our comparison of distributed and in person teams yields four interesting observations. First, distributed teams' conversations resulted in fewer clusters than the conversations of in person teams, suggesting that distributed teams may be more parsimonious in their discussions. This parsimony may be attributable to the limitations of the media through which they are communicating.

Second, in person teams may avail themselves of media richness. Specifically, they may use visual cues to orchestrate how work is allocated among team members. Unrestricted and time pressured teams had quartiles with no clusters that included operational strategy. In fact, time pressured teams had no clusters with operational strategy during the first three quartiles of the first performance episode. These observations suggest that in person teams may be employing a "just because you can" strategy. In other words, instead of exploiting the richness of in person communication by relying on visual cues, their overall performance and role clarity may be enhanced by streamlining conversations and clearly articulating an operational strategy, respectively.

Third, distributed teams with good cost performance may use *tactical strategy-clarify* to facilitate establishing and verifying their tactical strategies. In these teams, *tactical strategy-clarify* is initially clustered with other types of tactical strategy messages. During the second and/or third quartile, it shifts to being clustered with mission analysis. Finally, it returns to a tactical strategy cluster. This shifting may indicate that *tactical strategy-clarify* is used over time to help establish a reasonable tactical strategy, ensure the action adheres to that strategy, and finally verify that the solution aligns with it. Interestingly, the same pattern can be observed in unrestricted in person teams, but not in those with limited time available.

Finally, unrestricted distributed teams and time pressured in person teams have a cluster that includes *tactical strategy-inform* and *tactical strategy-suggest* (and sometimes *tactical strategy-agree*) in the second, third, and fourth quartiles during the first performance episode. The teams with good cost performance appear to use this cluster to help them verify the efficiency of their solution. In teams with bad cost performance, alternatively, this cluster may be an indicator of their struggles as they repeatedly offer alternative strategies for consideration. This pattern also was observed in unrestricted in person teams, particularly in those teams with bad cost performance. This cluster was not observed in distributed teams facing time pressure.

Comparisons of Unrestricted Versus Time Pressured Teams

Three interesting sequential patterns were apparent when comparing unrestricted and time pressured teams. First, teams with unrestricted time have the luxury of engaging in detailed conversations about mission analysis, as demonstrated by the clusters in the first quartiles focused solely on exchanging information about the mission analysis. The culmination of these conversations is a summary of the mission analysis that is clustered with *tactical strategy-summarize* (distributed teams only) and *operational strategy*. This combination suggests that teams with unrestricted time can simultaneously consider what, why, and how they are going to accomplish their task. Time pressured teams, may have to be satisfied with partial information before diving into action. The time pressured distributed teams demonstrate this scenario through the cluster *tactical strategy-clarify, tactical strategy-inform, operational strategy* and *mission analysis-summarize*, which suggests that they may not have a fully developed tactical strategy during the first quartile but are willing to apply what they do know as they plan next steps.

Second, the sequential patterns observed during the second performance episode differed between unrestricted and time pressured teams with good time performance. Unrestricted teams, regardless of media, engaged in conversations comprised of similar patterns during the first quartile of both performance episodes. Second quartile patterns, however, differed; where the sequential patterns during the second quartile of the second performance episode were more similar to the patterns exhibited during the third quartile of the first performance episode. In other words, unrestricted teams started both episodes similarly, but their experience with the task facilitated more quickly transitioning to focused attention on action. Alternatively, time pressured teams' first quartile during the second performance episode was more similar to their second quartile during the first performance episode, which suggests that they were able to use what they learned to give themselves a head start in the second performance episode.

Third, across all conditions, we observed that teams with bad cost performance struggled to obtain a common understanding of their mission and devise a tactical strategy. Later conversations were often focused on clarifying the parameters governing their mission or suggesting different approaches to complete the task. In other words, the teams repeatedly covered the same ground without any resolution or forward progress. Frustration was not uncommon in the transcripts. But, that frustration in the time pressured teams, often resulted in the teams giving up.

Theoretical Implications and Future Research

Our investigation of the sequential patterns that exist in team communication provides additional confirmation of Marks and colleagues (2001) recurring phase model of team processes and Bartunek and Woodman's (2015) temporal dimensions of change. Marks and colleagues (2001: 360) posit that "outcomes from initial episodes often become inputs for the next cycle." Our findings support this contention in two specific ways. First, the sequential patterns established during the first performance episode continued into the second. Teams with good and bad cost performance demonstrate this phenomenon. The good cost teams used what they learned in the first performance episode to expediently understand the similarities and differences in the task requirements during the second performance episode. Alternatively, the bad cost teams, who exhibited confusion and frustration during the first performance episode, continued to struggle during the second. Second, all good cost performance teams exhibited an expedited path toward completion during the second performance episode when compared to the first; the way in which they expedited their conversations, however, was dependent upon their time availability. Time pressured teams were able to skip over the preliminary conversations in which they engaged during the first quartile of the first performance episode. Instead the first quartile conversations during the second performance episode were very similar to the second quartile conversations of the first performance episode. Unrestricted teams had similar first quartiles across the two performance episodes, but the second quartile during the second performance episode was very similar to the third quartile in the first. Together these findings demonstrate how teams use what they have learned from one performance episode to the next, but that learning may not always be beneficial and how it influences the team communication may be dependent on the collaboration conditions.

When considering the temporal dimensions of change, Bartunek and Woodman (2015: 170–171) suggest that "some sequences associated with change involve considerable cycling back and forth." In our examination of the sequential patterns of team communication over time, we identified several recurring clusters. But, these clusters may not always serve the same purpose over time. For instance, we observed that *tactical strategy-clarify* changed clusters over time from clusters focused on tactical strategy to mission analysis and back to tactical strategy. This cycling back and forth facilitated the change required for the team to accomplish its task, thereby providing evidentiary support for Bartunek and Woodman's (2015) conceptualization. Our findings, however, extend this notion by suggesting that the functions of these repeating cycles may differ over time. For example, the repeated use of *tactical strategy-clarify* facilitated devising an initial tactical strategy, then ensuring that it was being applied, and finally verifying that the solution aligned with it.

Two additional theoretical contributions may be extrapolated from our findings. First, over the course of a performance episode, teams may need to engage in both divergent and convergent thinking, which are used to generate many ideas and then select the good one, respectively (Paletz and Schunn, 2010). The cognitive processes associated with divergent thinking, as defined by Paletz and Schunn (2010), are information search and analogy, whereas convergent thinking requires appropriate evaluation. These cognitive processes are seen throughout the teams' communication strings and this conceptualization, therefore, may help explain the differences we observed between teams with good and bad cost performance. The good performance teams appear to be able to switch between divergent and convergent thinking throughout the performance episodes. Specifically, their conversations contain sequential patterns that depict teams sharing information about their mission and the parameters associated with their unique roles, ensuring adherence to their tactical strategies, and evaluating the efficacy of their solutions vis-à-vis their mission parameters and tactical strategies. Alternatively, the bad performance teams continue to have clusters representing divergent thinking throughout the performance episodes. These teams may struggle with engaging in convergent thinking or identifying when they should make that shift. Either way, planning communication activities that incorporate both types of thinking may be imperative for achieving high levels of performance.

The second theoretical framework applicable to our research on sequential patterns is Klein et al.'s (2007, 2010) sensemaking, which is a reciprocal process where data is used to construct explanatory structures that are then used to identify relevant data. Sensemaking is useful to construct both a functional (what should we do) and an abstract (what might be done) understanding of the situations in which we find ourselves (Klein et al. 2007). This distinction may help explain the difference between the clusters of unrestricted and time pressured teams that contain operational strategy. Unrestricted teams' clusters include *tactical strategy-summarize*, *mission analysis-summarize* and *operational strategy*. These teams appear to work toward a functional understanding of both their mission analysis and tactical strategy such that they can summarize each as they determine how they will allocate work. In other words, they employ a functional understanding of mission analysis and tactical strategy as they plan their operational strategy. Alternatively, time pressured teams' cluster replaces the *tactical strategy-summarize* with *tactical strategy-clarify, tactical strategy-inform*. Given the time pressure under which they are functioning, they move forward with a more abstract understanding of what their tactical strategy might be. Abstract functioning, however, may not always be preferable to a team, even it if would help them progress on their task. For example, the teams with lower cost performance exhibited communication behaviors suggesting that they were less able to collaborate effectively with an abstract understanding, particularly about their mission analysis. Therefore, training team members about the value of abstract understanding may be useful.

At least two opportunities for future research exist to build upon our findings in this chapter. First, adding a second dimension (i.e., purposes) to our examination of longitudinal team communication contributes to the literature. At the same time, it also exposes an opportunity for research. To better understand why the good and bad cost teams had sequential pattern similarities, we looked to the transcripts and found that the specific content exchanged differed within these sequential patterns. These insights suggest that more research focused on the specific information team members exchange is warranted. Second, as we recommended in Chap. 5, an analogous examination of the specific action processes delineated by Marks and colleagues (2001) may be fruitful. These processes are monitoring progress toward goals, systems monitoring, team monitoring and backup behavior, and coordination activities. Moreover, since some of these processes may not require communication, this type of investigation could combine team communication with physical actions to garner a more thorough understanding of how teams accomplish their tasks.

Methodological Contributions and Future Research

Our efforts extend the work of several researchers interested in longitudinal human behavior. First, Olson and colleagues (1994) visualized the temporal structure of team meeting conversations using hierarchical cluster analysis. Their overarching purpose was to demonstrate the applicability of this technique to longitudinal data rather than the theoretical implications. Our efforts, conversely, focus on applying the technique to ascertain theoretical implications. Second, Tschan and Waller are two researchers who have examined patterns over time. The work of Waller and colleagues, in particular, demonstrates that the timing of behavioral patterns may link to performance (e.g., Uitdewilligen et al. 2018; Waller 1999; Wang et al. 2016). For example, Waller (1999) studied the patterns present in three behaviors as teams adapted to nonroutine events and found that performance was impacted more by the timing of behaviors than the frequency of their occurrence. Tschan (1995, 2002) has investigated communication cycles and found that higher quality cycles are exhibited by higher performers at both the individual- and group-levels. Our research combines and extends these lines of inquiry by examining team communication patterns over time, including the quality of what has been said, to ascertain any temporal structures. We extend these efforts by investigating how the sequential patterns differed over time, across the conditions under which teams were working, and between good and bad performing teams.

Our methodological approach also includes visualizations. Less than half of the papers we reviewed in Chap. 2 include visualizations of their data and/or results. Visualizations, particularly when multiple comparisons are desired, provide an opportunity to identify trends, similarities, differences, anomalies, et cetera. For example, the visualizations facilitated our identifying the *tactical strategy-inform* and *tactical strategy-suggest* sub-clusters across multiple quartiles and team types. We may not have investigated them had we only reviewed results depicting the clusters with distances greater than 6. But, they provided one perspective on our finding that teams with good and bad cost performance have similar clusters but dissimilar results, which may be explainable by the specific content of their communication.

Managerial Minutes

At least some time pressure is a good thing.
Teams with minimal time restrictions may need help maintaining focus and moving forward. You may want to consider giving all teams at least some time pressure to help drive them toward their deadlines.

Ask questions to clarify how the team is approaching its task.
At the beginning questions can help hone the team's strategy, next they can be used to make sure the strategy is being followed as the work is done, and finally they can help verify that the team did what it planned to do and achieved the anticipated result. Identifying someone willing to take on this role for the team may be advantageous.

Asking questions can also be detrimental.
You need to keep an eye on what is being asked. If too many questions focus on rehashing what you thought was already decided, the team may be heading toward bad performance. If the entire team is struggling, you may want to dedicate a meeting to getting the team back on track. If individual team members seem to be struggling, you may want to work with them offline to help them get up to speed.

Under time pressure, help teams iterate and move forward.
Time pressured teams may need help moving forward without complete information. When time is particularly tight, you may want to assign individuals who are more comfortable working with ambiguity.

Future researchers may find value in looking at longer or more distant transitions in the data. The $t + 1$ transitions are a viable starting point, but as our clusters demonstrate, much more complex sequential patterns may exist. Additionally, our research could be extended by broadening the dataset to include physical behaviors, gestures, artifacts, or other variables that might represent the collaborative process. This type of extension would be particularly appropriate for those interested in examining the action processes of teams.

Appendix: How to Conduct Cluster Analysis and Construct Dendrograms

In Chap. 7, we examined the sequence of communication by using hierarchical clustering to identify sequential structures in the communication strings. We then used dendrograms to provide a visual representation of the clustering. To demonstrate these methods, we evaluate the example dataset presented in the Chap. 3 Appendix. The pattern recognition and measurement, as well as the visualization construction, were conducted in Matlab, while the statistical analyses were completed in Excel.

Data Preparation

In this analysis, we dive deeper into the communicated message by not only looking at what is said but how it is said. That is, we include both topic and purpose codes in the example dataset. In our coding system, the topic level is the tens digit and the purpose is the ones digit in the code. For example, a message coded as 13 means that the message is about topic 10 and the purpose for delivering the message was 3.

We prepared our data by calculating the frequency of topic-purpose transitions in the dataset. Specifically, we counted the number of times the teams transitioned between every possible combination of topic-purpose codes (e.g., transition 11 to 12 or 11/12). For example, Team 1 in our example dataset has the communication string:

11, 12, 11, 21, 23, 23, 23, 31, 32, 31, 12, 13, 12, 41, 41, 42, 41, 41, 41, 41.

For this team, the counts of transition from 11 to all other codes would be:

11/11 = 0, 11/12 = 1, 11/13 = 0, 11/21 = 1, 11/22 = 0, 11/23 = 0, 11/31 = 0, 11/32 = 0, 11/33 = 0, 11/41 = 0, 11/42 = 0, and 11/43 = 0.

These counts are summed with the counts from all other teams in the dataset. The total frequency counts for the entire sample were compiled in a matrix, which is depicted in the top of Table 7.2.

Table 7.2 Actual and expected transitions in example dataset

Transitions in the example dataset															
		To Topic-Purpose													
		11	12	13	21	22	23	31	32	33	41	42	43	Totals	Actual Relative Frequency
From Topic-Purpose	11	4	15	0	1	2	3	2	1	0	3	3	0	34	0.16
	12	15	5	6	0	1	0	3	0	0	2	0	0	32	0.15
	13	3	2	0	0	0	0	0	0	0	1	0	0	6	0.03
	21	0	0	0	0	1	1	0	1	1	1	1	0	6	0.03
	22	2	0	0	4	5	6	2	1	0	1	2	0	23	0.11
	23	0	0	0	1	7	3	2	0	0	1	0	0	14	0.07
	31	1	4	0	0	1	0	1	13	0	2	0	0	22	0.1
	32	0	3	0	0	1	0	11	3	1	1	1	0	21	0.1
	33	1	1	0	0	0	0	0	0	0	0	0	0	2	0.01
	41	4	0	0	0	2	0	0	2	0	10	11	0	29	0.14
	42	1	0	0	0	1	1	0	0	0	10	8	0	21	0.1
	43	0	0	0	0	0	0	0	0	0	0	0	0	0	0
Totals		31	30	6	6	21	14	21	21	2	32	26	0	210	1
Actual relative frequency		0.15	0.14	0.03	0.03	0.1	0.07	0.1	0.1	0.01	0.15	0.12	0	1	

Expected transitions														
		To Topic-Purpose												
		11	12	13	21	22	23	31	32	33	41	42	43	Totals
From Topic-Purpose	11	5.02	4.86	0.97	0.97	3.40	2.27	3.40	3.40	0.32	5.18	4.21	0	34
	12	4.72	4.57	0.91	0.91	3.20	2.13	3.20	3.20	0.30	4.88	3.96	0	32
	13	0.89	0.86	0.17	0.17	0.60	0.40	0.60	0.60	0.06	0.91	0.74	0	6
	21	0.89	0.86	0.17	0.17	0.60	0.40	0.60	0.60	0.06	0.91	0.74	0	6
	22	3.40	3.29	0.66	0.66	2.30	1.53	2.30	2.30	0.22	6.50	2.85	0	23
	23	2.07	2.00	0.40	0.40	1.40	0.93	1.40	1.40	0.13	2.13	1.73	0	14
	31	3.25	3.14	0.63	0.63	2.20	1.47	2.20	2.20	0.21	3.35	2.72	0	22
	32	3.10	3.00	0.60	0.60	2.10	1.40	2.10	2.10	0.20	3.20	2.60	0	21
	33	0.30	0.29	0.06	0.06	0.20	0.13	0.20	0.20	0.02	0.30	0.25	0	2
	41	4.28	4.14	0.83	0.83	2.90	1.93	2.90	2.90	0.28	4.42	3.59	0	29
	42	3.10	3.00	0.60	0.60	2.40	1.40	2.10	2.10	0.20	3.20	2.60	0	21
	43	0	0	0	0	0	0	0	0	0	0	0	0	0
Totals		31	30	6	6	21	14	21	21	2	32	26	0	210

Data Usage

Chi-Squared Test of Independence

The matrix of transition data was assessed against a matrix of expected values. In Table 7.2, the data from the example dataset is in the top matrix and the expected values are in the bottom matrix. The top matrix also includes the actual relative frequencies, which are calculated by dividing the total number of transitions in the given row or column by the total number of all transitions. For example, in the first row of Table 7.2, 34 transitions started with an 11 code out of the 210 transitions exchanged, which translates into an actual relative frequency of 34/210 = 0.16.

The expected value for a specific cell is calculated by multiplying the total number of transitions by the actual relative frequencies of the row and column corresponding to that cell. For example, the expected value of an 11 to 11 transition is $210 \times 0.16 \times 0.15 = 5.02$. Note that some slight differences in the thousandths digit are attributable to rounding error.

Those topic-purpose transitions with near zero or zero values were omitted. The rationale for omissions is to ensure that chi-squared test assumptions are not violated and to be parsimonious. For the example dataset, all transitions including 43 were omitted because 43 never occurred in the dataset. We also removed all transitions including 13, 21, 33, and 42 because their expected values were less than 1.

Table 7.3 Resulting actual and expected transitions in example dataset needed to run chi-squared test of independence

Transitions from example dataset											
		To Topic-Purpose									
		11	12	21	22	31	32	41	42	Totals	Actual relative frequency
From Topic-Purpose	11	4	15	1	2	2	1	3	3	31	0.18
	12	15	5	0	1	3	0	2	0	26	0.15
	21	0	0	0	1	0	1	1	1	4	0.02
	22	2	0	4	5	2	1	1	2	17	0.10
	31	1	4	0	1	1	13	2	0	22	0.13
	32	0	3	0	1	11	3	1	1	20	0.11
	41	4	0	0	2	0	2	10	11	29	0.17
	42	1	0	0	1	0	0	10	8	20	0.12
Totals		27	27	5	14	19	21	30	26	169	1
Actual relative frequency		0.16	0.16	0.03	0.08	0.11	0.12	0.18	0.15	1	

Expected transitions										
		To Topic-Purpose								
		11	12	21	22	31	32	41	42	Totals
From Topic-Purpose	11	4.95	4.95	0.92	2.57	3.49	3.85	5.50	4.77	31
	12	4.15	4.15	0.77	2.15	2.92	3.23	4.62	4.00	26
	21	0.64	0.64	0.12	0.33	0.45	0.50	0.71	0.62	4
	22	2.72	2.72	0.50	1.41	1.91	2.11	3.02	2.62	17
	31	3.61	3.61	0.65	1.82	2.47	2.72	3.91	3.38	22
	32	3.20	3.20	0.59	1.66	2.25	2.49	3.55	3.08	20
	41	4.63	4.63	0.86	2.40	3.26	2.630	5.15	4.46	29
	42	3.20	3.20	0.59	1.66	2.25	2.49	3.55	3.08	20
Totals		27	27	5	14	19	21	30	26	169

The resulting matrix (provided in Table 7.3) was then analyzed using a chi-square test of independence to determine if the dataset of transitions was different than a random dataset. The chi-square test indicates that significant structure exists in the dataset ($\chi 2 = 248.1$, $p < 0.0001$).

Hierarchical Clustering

To better visualize any sequential structures in the topic-purpose transitions, we use hierarchical clustering. The clustering is determined via the Euclidean distance between the topic-purpose transitions within the data. That is, using the Euclidean distance we can identify which topic-purpose transitions are closest (lowest distance) to another. When these connections are plotted in dendrograms we see a treelike structure where those connections with the lowest distance are clustered near one another, and those with larger distances are further apart. Figure 7.3 shows the dendrogram representation of the example dataset.

The final step in analysis is to determine a cutoff value that separates clusters for discussion purposes. For example, in the dendrogram depicting the example dataset if the cutoff was set at 3.5, 4 clusters would be available for interpretation: (1) codes 12 and 11, (2) codes 31 and 32, (3) codes 22 and 21, and (4) codes 42 and 41.

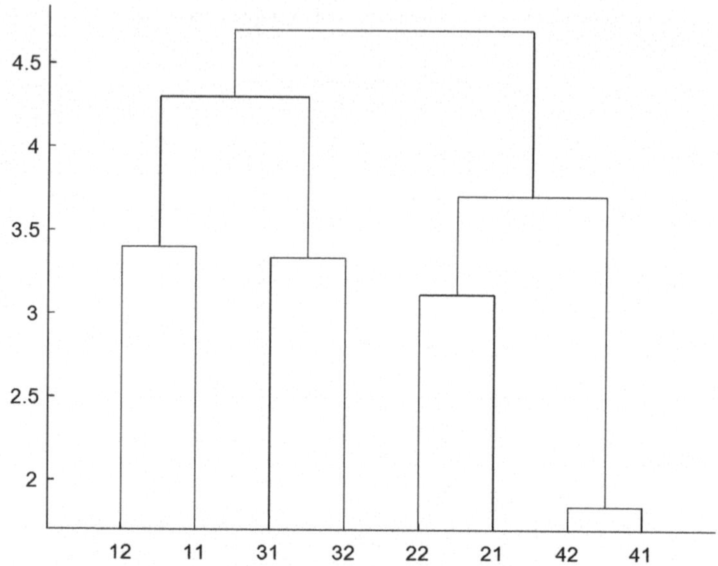

Fig. 7.3 Dendrogram of all teams in example dataset

However, if the cutoff was set at 4.0 then three clusters would be interpreted: (1) codes 12 and 11, (2) codes 31 and 32, and (3) codes 22, 21, 42, and 41.

The determination of the cutoff is a researcher's decision that balances parsimony and conceptual value in the interpretation of clusters. That is, on the one hand, little value may come from discussing each code as its own cluster. On the other hand, considering everything as one cluster may provide little insight. Thus, a researcher has to use judgment in selecting a cutoff value for clusters that facilitates interpreting a reasonable number of clusters that are of practical and/or theoretical interest.

Cross-Condition Comparisons

Hierarchical clustering analysis using dendrogram representations also lends itself to visual comparisons across conditions. For example, we split the dataset at the performance median in Chap. 7 to gain greater perspective about the way teams utilize topic-purpose statements. The split was done by finding the median performance of all teams in a given condition, then grouping those teams above the median into a "good" condition and those teams below the median in the "bad" condition.

To demonstrate with our example dataset, the median performance score is 0.525. As such, Teams 1, 2, 3, 4, and 6 are above the median with performance scores of 0.98, 0.85, 0.75. 0.90, and 0.65 respectively, and therefore in the "good condition." Teams 5, 7, 8, 9, and 10 are below the median with performance scores of 0.40, 0.30, 0.40, 0.22, and 0.30 respectively, and therefore in the "bad condition."

Using the methods described above, we constructed matrices of the transitions data for both conditions and conducted a chi-squared test of independence. We found that both matrices are significantly different than expected ($p < 0.0001$). Next, hierarchical clustering analysis was applied to each condition and the dendrograms were created for good (see Fig. 7.4a) and bad (see Fig. 7.4b) performing teams.

In determining the cutoff to apply when interpreting the dendrograms, we considered cutoffs of 2, 2.5 and 3. At a cutoff of 2.0 all topic-purpose codes are their own clusters except 41 and 42, and these cluster together in both conditions. At a cutoff of 2.5 bad teams wrap in 22 with the 41–42 cluster, yet otherwise topic-purposes are their own clusters. At a cutoff of 3.0, we see fewer individual clusters and more differences across clusters, which may make for a meaningful comparison. Notably, at a 3.0 cutoff, good performing teams have five clusters (i.e., 22–23, 31, 32, 11–12, and 41–42) whereas the bad performing teams have four clusters (i.e., 31, 32, 11–12, and 41–42–22). As previously mentioned, deciding on the most appropriate cutoff is based on the researcher's judgment.

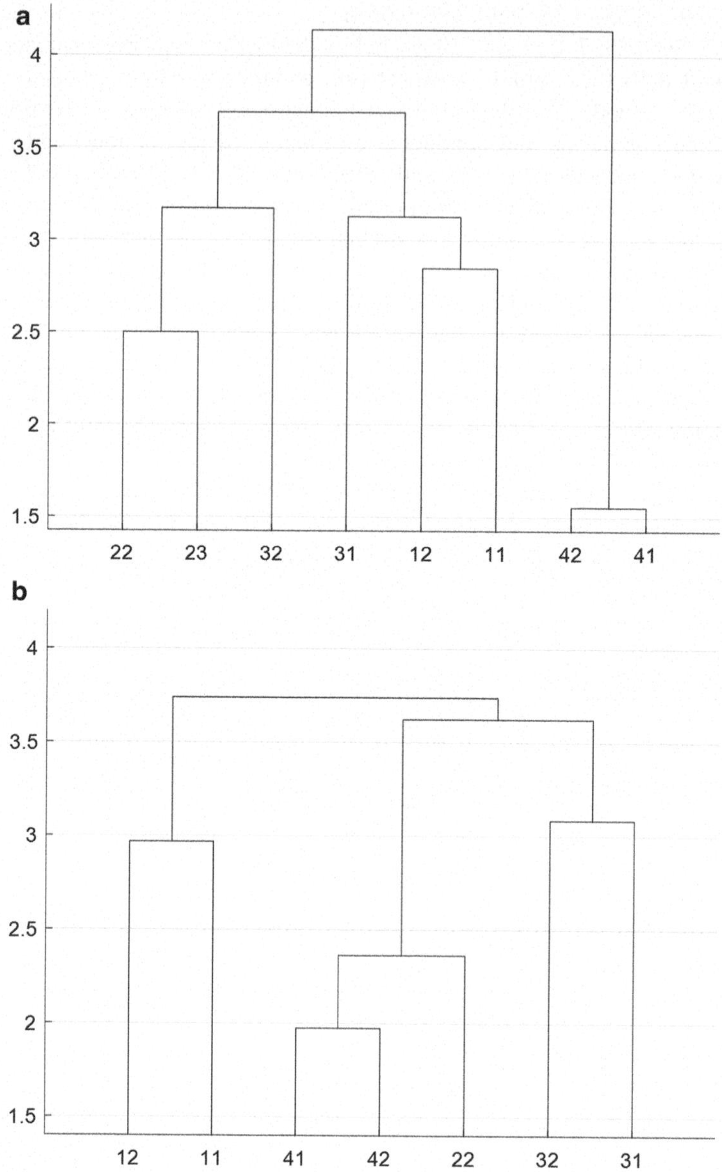

Fig. 7.4 Dendrograms of good and bad performing teams in example dataset. (a) Dendrogram of good performing teams from example dataset, (b) Dendrogram of bad performing teams from example dataset

Chapter 8
What We Learned about Longitudinal Team Communication

Organizations often utilize teams to provide competitive advantage through the interdisciplinary perspectives of the members. Yet, to achieve their organizational goals, teams must collaborate and coordinate activities throughout their time together. Understanding and managing teams, therefore, may be enhanced by examining the way activities unfold over time. We engaged in such an examination by focusing on longitudinal team communication because communication provides a glimpse into the inner workings of teams. The theoretical foundations of our study are the recurring phase model of team processes developed by Marks et al. (2001) and different temporal dimensions of change forwarded by Bartunek and Woodman (2015).

The recurring phase model characterizes team activity as an iterative series of phases within performance episodes comprised of (1) planning and evaluative activities, called transition processes and (2) the actual work being done, called action processes (Marks et al. 2001). Specifically, we captured when teams discuss these transition and action processes and identified patterns in their conversations. The transition processes include mission analysis, goal specification, and strategy formulation and planning. Based on our assessment of the latter, we discovered that teams used formulated plans about tactical strategy and operational strategy differently and, therefore, examined these processes separately in our analyses. Given the task completed by our teams, we determined that a singular focus on action was appropriate, so all action processes are examined together. This focus also allowed us to isolate the intricacies associated with how teams use the various transition processes over time.

Bartunek and Woodman's (2015) temporal dimensions of change guided our examination of team communication, as we focused on the temporal changes in transition and action processes across performance episodes. Their dimensions characterize dynamic, non-linear changes over time as having rhythm, timing, pacing, polyphony, and sequence. Through our research, we garnered insights about longitudinal team communication along these dimensions of change. First, the

S. McComb, D. Kennedy, *Computational Methods to Examine Team Communication*, Computational Social Sciences,
https://doi.org/10.1007/978-3-030-36159-4_8

rhythm of communication was found to include cyclical discussions that shift among topics over time (Chap. 4). Second, timing and pacing were examined together in Chap. 5. Timing reveals changes in how frequently teams communicated about different transition processes or action processes over the four quartiles in two performance episodes. Pacing shows that the number of topics discussed reduces over time indicating a progression from planning to action as teams work on their tasks. Third, polyphonic changes in communication patterns elucidate how topics interrelate and may prompt cyclical conversations to happen over time (Chap. 6). Finally, our examination of sequence shows that in addition to understanding what topics are broached, the way in which these topics are discussed may influence a team's ability to be effective (Chap. 7).

By interweaving these frameworks, we provide insights into what, when, how, and why team communication changes over time, including any corresponding impact on performance. Further, we showed that conditions under which teams are collaborating matter. That is, whether the team was distributed and using a collaboration medium to communicate versus working face-to-face, and if the team was unrestricted in the amount of time available to complete its task or working under time pressure, affected communication patterns. We now bring these insights together to look holistically into what has been learned from this research. In particular, we translate the research findings into practical takeaways to show how such a thorough examination culminates in specific research recommendations and on-the-ground managerial implications. We then articulate the theoretical contributions that may be derived from our efforts and the value of expanding team researchers' methodological toolboxes. While acknowledging the limitations of this study, we advance opportunities for new research in this area. We end with concluding thoughts about longitudinal team communication and the importance of considering it to be an integral component of a team's journey worthy of focused consideration.

Insights Garnered From a Holistic Examination of Longitudinal Team Communication

In this section, we demonstrate the way in which our work advances holistic insights about team communication. Specifically, we amalgamate the results and implications from all of our analyses to distill five key themes about team communication, including the impact of time availability, media used, and experience working across multiple performance episodes. These themes draw together multiple insights gained from our investigation and provide the impetus for practical implications. They are:

1. Engage in transition processes before action processes
2. Recognize the changing relevance of transition processes over time
3. Avoid the curse of unrestricted time
4. Exploit learning over time
5. Satisfice under time pressure

Engage in Transition Processes Before Action Processes

When teams work together on a task, they engage in a number of processes as they understand and plan their work (transition processes), as well as perform the work (action processes). The recurring phase model (Marks et al. 2001) suggests that, over time, teams will cycle between transition and action processes. Our research confirms this suggestion. Moreover, our results indicate that teams, in general, start with transition processes and then shift to action processes. We find evidence of this progression from our examination of how, how much, and why topics are being discussed in both performance episodes.

In Chap. 4, the recurrence analysis results suggest that teams are focusing on different topics over time, as can be visualized in the checkboard structure of recurrence plots and inferred from the non-zero recurrence quantification metrics. In Chap. 5, we focused on the timing of topic discussions and found that messages about mission analysis and tactical strategy transition processes were most frequent in earlier quartiles, with later quartiles showing higher frequencies of action. Yet, looking deeper into the interrelationships among these processes shows that conversational shifts are not, and probably should not be, conclusive; rather, the conversational shifts are iterative. In Chap. 6 for instance, we found significant interrelationships between when transition processes and action occur, thereby suggesting that teams may iteratively test the feasibility of their plans by testing them through action. This idea is further substantiated by the investigation undertaken in Chap. 7, where we examine sequential patterns in the message-to-message exchanges. Our results suggest that over time mission analysis, tactical strategy, and operational strategy clusters become less prominent and more distant from the cluster containing action. An examination of the team transcripts demonstrates that these sequential patterns depict a shift from focused efforts on transition processes with iterative bursts of action assessing feasibility to more focused action with short bursts of checking work against established plans.

For teams in practice this finding may provide insights for organizing and managing work. That is, team meetings could be organized to ensure this progression is followed. Earlier meetings focus on (1) ensuring continuity in how members understand their mission and the parameters governing them and (2) establishing an approach for conducting the work. As teams garner consensus regarding their transition processes, meeting time can progressively shift toward more updates on taskwork and confirmations that progress is adhering to the established plans. Throughout this progression, managers must remember the need to keep iterating between transition processes and action processes; this continued connection between the two types of processes enables teams to check their work. Indeed, if teams are not iterating, managers may need to interject and encourage teams to revisit their assumptions and decisions to ensure they remain viable, alter them if necessary, and confirm that their actions continue to align with their parameters and plans.

Recognize the Changing Relevance of Transition Processes Over Time

The previous theme focused on the relationship between transition and action processes. This theme dives into the relationships among the transition processes. More specifically, our results show that these processes are relevant at different times throughout the team's lifecycle, yet in an ordered fashion and for different reasons. Nevertheless, our most troubling finding suggests that teams with seemingly similar communication patterns among the transition processes achieve different levels of performance, thereby underscoring the need to monitor both *when* and *why* teams are engaging in conversations about transition processes.

In Chap. 4, we see broken recurrence analysis boxes in plots, which imply that teams are interjecting topics into otherwise focused conversations. Moreover, during the second performance episode, we observed more periodic bursts across the recurrence plots. From Chap. 5, we find that, in general, mission analysis precedes tactical strategy. That is, the frequency with which teams discuss mission analysis declines from the first to the second quartile, whereas tactical strategy declines from the second to the third. This pattern continues into the second episode, although with overall fewer messages exchanged. Teams appeared to find operational strategy more relevant during the second performance episode.

In Chap. 6, we note that the number of conversations about mission analysis affects the number of conversations about tactical strategy and vice versa. This interconnection may be detrimental to time performance if the teams iterate among these two transition processes too frequently. Simultaneously, an earlier initial conversation about tactical strategy may enhance cost performance. Finally, from Chap. 7, we see that the iterations among transition processes are only part of the story; how the teams are using these conversations may be more important to understand. Indeed, we saw similar sequential patterns between the high and low performing teams, but when we reviewed the transcripts, differences were observed. Specifically, higher performing teams engaged in purposeful conversations that moved them forward. These conversations helped teams check assumptions, build shared understanding, and, when necessary, modify what they were doing to better meet the needs of the task. Alternatively, the iterative exchanges among the members of lower performing teams demonstrated continued confusion that they were trying to resolve.

Teams, therefore, may benefit from starting with discussions about what they have been tasked with accomplishing. Next, they can use the collective understanding gained from those conversations to devise an approach by iterating between mission analysis and tactical strategy. Once they have shifted into their focus on action, periodic bursts of conversation about mission analysis, operational strategy, and tactical strategy are warranted. Interesting, this progression comes with two caveats. First, iterating between mission analysis and tactical strategy is important, however, too many iterations may adversely impact a team's time performance. More research is needed to determine how much iteration is appropriate and how to effectively use bursts of conversation to maintain forward progress without adversely impacting performance. Second, shifting among transition processes may signal that a team is high function *or* that a team is floundering. Without including the purpose for exchanging each message, we would not have uncovered this interesting distinction. Future researchers may benefit from focusing on how to distinguish between functional and dysfunctional iterative cycles and devising interventions to help teams caught in dysfunctional spirals to recover.

In practices, many teams start off by assigning work and setting up a time to reconvene for a status update. Our results strongly suggest that this process is not desirable. Instead, teams need to be encouraged to clarify their missions, including what is required and any corresponding parameters, before establishing an approach based on those boundaries and constraints. Once a plausible tactical strategy is devised, teams can then delineate who will do what by formulating their operational strategy. These conversations, however, should be monitored to ensure teams move forward effectively. In particular, teams may want to engage in periodic clarification; initially, they can ensure that the strategy devised aligns with their mission analysis, then question if it is being applied appropriately as they engage in action processes, and finally verify that it produced the anticipated result. Teams continually revisiting their mission analysis or tactical strategy to ask questions or gather more information, may be having trouble finding, or agreeing upon, a viable strategy that works within their task parameters. Indeed, repeated cycles may suggest that the processes are out of control, and teams are spiraling toward poor performance. As such, managers may need to step in and help the team iterate through the sequence again to establish an understanding of the boundaries and constraints, and build a plan that aligns with that understanding.

Avoid the Curse of Unrestricted Time

Our investigation captured teams working under different conditions. The findings around teams with unrestricted time offer an interesting insight about how beneficial time constraints might be for teams. In Chap. 4 we see that teams without time pressure engage in uniform, predictable patterns of communication during the second performance episodes, whereas time pressured teams engage in more random, dynamic conversations. This difference may indicate that teams with unrestricted time are less agile. In Chap. 5, we see unrestricted teams exchange more messages overall. Further, the distributed teams with unrestricted time exhibit more complex patterns of exchange among transition processes. In other words, they are bouncing among topics more frequently than the other types of teams. In Chap. 7, we found that unrestricted teams' conversations have more clusters. Interestingly, in person teams never discuss operational strategy, which may indicate that they are exploiting visual cues available to them, but may be lacking a precise understanding about who is doing what. The sequential patterns of communication from one performance episode to the next are also interesting. Unrestricted teams seem to engage in the same pattern of conversation in the first quartile of both performance episodes. Alternatively, time pressured teams are able to start the first quartile of the second performance episode with conversations that are more similar to the conversations in which they engaged during the second quartile of the first performance episode. In other words, time pressured teams appear to leap frog over the initial conversations to get into action more quickly after they have experience with the task.

Unrestricted time may, at first glance, be considered a luxury because teams can conduct their work without concern for elapsed time. But, our findings suggest that it may, in fact, be a drawback because it facilitates a wandering focus and more lengthy processes. In comparison, the patterns of communication exhibited by time pressured teams appear to be more efficient, in that these teams engaged in fewer exchanges, jumped into action more expeditiously, and articulated an operational strategy. Very little research focuses on the implications of time pressure on teams (see brief review in Chap. 1). Our results, however, demonstrate the need for such inquiry; and researchers may want to focus on the patterns we identified. First, a more comprehensive understanding about how time pressured teams streamline their exposition is imperative. In our study, distributed teams with unrestricted time appeared to struggle the most, thereby suggesting that the combination of no visual cues and no time constraints may be the most difficult environment in which to effectively communicate. Therefore, this line of inquiry is particularly important for teams working via distributed communication media. Second, the ability to quickly engage in action, particularly in subsequent episodes, implies that team members are applying lessons learned from previous activities; we examine this aspect of our results more thoroughly in a subsequent theme. Finally, the balance between the use of visual cues and the articulation of an operational strategy requires investigation. Indeed, the in person teams in our study could see that different task parameters may be present because each member was given their assignment on different colored paper, which may have prompted more discussions about transition processes. Yet, when these teams had unrestricted time, they did not engage in conversations about operational strategy, which may have limited team member role clarity and required more time to conclude team activities. Contrasting this behavior to that of time pressured teams, with a particular focus on the relative benefits of visual cues and operational strategy formulation, warrants attention.

In reality, teams without at least some time pressure are rare. Nevertheless, when an abundance of time is available, managers may need strategies to help teams function efficiently. First, these teams may engage in more complex conversations, but the conversations may stray; managers, therefore, need to be ready to help teams get their conversations back on track. One plausible solution for when team members raise important, but off topic, points is to use a "parking lot" where these points can be stored and addressed later instead of diverting the current, incomplete conversation. Second, managers may benefit from identifying individual team members who can summarize conversations and help move the team toward consensus. These types of team members may help avert unfocused conversations and an inability to make decisions. Finally, when teams have to work together over multiple performance episodes, engaging in a brief conversation about lessons learned from previous efforts may help the team focus more quickly on how they can use those lessons to launch into action.

Exploit Learning Over Time

We investigated team communication across two performance episodes and identified distinct changes in the communication patterns across them. In Chap. 4, the rhythm of the conversations decreases in complexity from the first to the second performance episode. Further, the recurrence plots suggest that teams move through cycles more rapidly in second performance episode. In Chap. 5, while we still see iterative cycles between transition processes and action, we note that they are less frequent during the second performance episode. While these findings demonstrate differences between the two performance episodes, the results from Chapters 6 and 7 provide insight into how teams may use learning across the performance episodes. The Chap. 6 results suggest that time pressure teams engage in their initial discussions about mission analysis and operational strategy earlier in the second performance episode. Moreover, they have fewer conversations about mission analysis and tactical strategy during the second performance episode. In Chap. 7, our results depict fewer clusters in the second performance episode than the first, from which we can infer that conversations about transition processes will be short and infrequent because the focus is on action. A review of the transcripts to interpret the Chap. 7 results, also indicates that teams with bad cost performance who floundered through ineffectual iterative conversations during the first performance episode, continue this pattern of dysfunctional behavior into the second performance episode.

Taken together these findings suggest that learning may occur when teams tackle multiple similar tasks. First, given the strong similarity between the tasks completed during both performance episodes, effective teams quickly ascertain the differences, assess their previous strategy given these changes, and jump right into action. In other words, they use what they learned during the first performance episode to streamline their planning processes during the second. This pattern of behavior was particularly relevant in teams facing time pressure. Second, bad behaviors can also be learned, as demonstrated by our teams who struggled. An area that has received little research attention is when teams are assigned repetitive tasks across multiple performance episodes. Our findings, however, demonstrate the potential for this line of inquiry. Several interesting opportunities exist including investigating how constructing a tactical strategy in the first performance episode parlays into performance at later episodes, how much similarity between tasks is needed to garner the learning demonstrated by our teams, how long teams can be dormant before learning effects are neutralized, and whether feedback about performance would enhance or diminish learning.

For practitioners, teams need to be encouraged to exploit what they have learned and apply it to future undertakings. This encouragement, however, may differ somewhat depending on the time constraints and the quality of initial work. Indeed, our results suggest that learning may be most relevant for time pressured teams and teams who struggle during their initial session. Managers, therefore, may need to help teams, particularly those facing time pressure, reach consensus about their transition processes and then rely on those decisions as they progress and initiate the next task. If teams were dysfunctional during their initial task, significant intervention may be required to help them unlearn bad communication behaviors and function more effectively as a team. Alternatively, managers may try mitigating dysfunction in the first place by facilitating team discussions that pattern efficient communication.

Satisfice Under Time Pressure

As we look at teams working under different conditions, the effect of time pressure appears to be more influential than the media available to the team. In Chap. 4, the recurrence rates for time pressured teams were similar regardless of media, whereas recurrence rate differences exist across media for teams with unrestricted time. Also, time pressured teams require different communication patterns across the two performance episodes to realize better cost performance; in the first performance episode teams may exhibit more predictable patterns whereas in the second performance episode teams may engage in more random and dynamic conversations. In Chap. 6, teams under time pressure were more succinct, having fewer meaningful discussions about tactical strategy and early conversations about mission analysis and operational strategy. In Chap. 7, teams under time pressure may not have had time to fully form their tactical strategy before making assignments. This inference is based on their tactical strategy cluster being comprised of messages that clarify and inform versus the tactical strategy cluster of teams with unrestricted time that contains summary statements. In other words, time pressured teams may have to move forward with a partial understanding that may not be formulated enough to summarize. This need to satisfice might overwhelm bad cost performing teams because they tend to give up on constructing a workable strategy.

Time pressured teams need to be both efficient and agile in their communication, which may require satisficing. They may compensate for satisficing, particularly during the second performance episode, by interjecting short bursts of communication about transition processes to make any real-time adjustments and ensure their action is on track. A more thorough investigation into how time pressured teams employ parsimony, agility, and satisficing is needed to better understand how their communication is structured. Strategies derived from such investigations would be helpful for all teams, particularly those teams who appeared to struggle with satisficing. This struggle was evident in the lower performing teams in our study who exchanged messages about the futility of planning and suggested completing their task through random actions. One plausible explanation for this struggle, which requires research, is that team members' personalities and/or knowledge, skills, and abilities may influence how effectively they can proceed with less than perfect information.

In organizations where teams are routinely time pressured, training opportunities focused on effective satisficing may be useful. Through such training, team members could learn how to (1) determine when they have enough information to effectively initiate action and (2) engage in purposeful, iterative conversations about transition processes that help the team hone their tactical strategy over time. As teams are collaborating, managers need to watch for team members frustrated by having to move forward with "too many" unknowns. Triggers may include attempts to fully articulate a tactical strategy, repeated requests for clarification, or comments about the entire effort being futile. When such triggers are identified, teams may require interventions that focus on what they have accomplished and how it informs their future actions.

Theoretical and Methodological Contributions From Our Examination of Longitudinal Team Communication

In addition to the five themes that were derived from a holistic view of our research findings, our efforts also make specific contributions that are worth noting. Table 8.1 summarizes these theoretical and methodological contributions of our research. For each of Bartunek and Woodman's (2015) temporal dimensions of change, our findings provide particular insights about how time plays a role in team conversations and that, when tracked may signal opportunities for intervention. As well, the methodological approaches, including both analytical and visualization techniques, are documented in the table. The choices we made were driven by our efforts to uncover nuances within teams' communication from the various temporal perspectives.

Theoretical Contributions Vis-à-Vis Bartunek and Woodman (2015) and Marks et al. (2001)

As we demonstrate in Table 8.1, our efforts have resulted in many theoretical connections and contributions. When viewed in totality, our most significant contributions are to the two theoretical frameworks underlying our research: Bartunek and Woodman's (2015) temporal dimensions of organizational change and Marks and colleagues (2001) recurring phase model. In this section, we highlight our contributions to these frameworks.

While Bartunek and Woodman's (2015) dimensions provide an organizing framework for our research, our results also demonstrate the applicability of these dimensions at the team level. Through our findings we observed differences in

Table 8.1 Summary of theoretical and methodological contributions

Temporal dimension	Data and method	Theoretical contribution	Methodological contribution
Rhythm of change – Repetitive cycles of change	*Data*: Topic communication strings. *Method*: Recurrence quantification analysis, linear mixed modeling, and recurrence plots.	We studied shifts among team processes and rhythmic repetition over time. The results provide evidence that teams cycle among transition processes and do so with different effectiveness; recurrent or iterative discussions on topics, focused or deterministic attention, and chaos or change among topics can affect performance. Results further confirm Bartuek and Woodman's (2015) position that change does not occur linearly as suggested by Lewin (1947). Helping teams determine when rhythmic repetition may be useful is an opportunity for intervention.	We applied recurrence quantification analysis to assess the metrics of recurrence rate, determinism, and entropy. These metrics were evaluated in linear mixed models. This approach uses longitudinally scored metrics to reveal variability in change over time. Recurrence plots of the rhythmic patterns over time were created, which helps visualize team progression.
Timing – When events occur, including the "right" time	*Data*: Topic frequency counts by quartile. *Method*: Statistical comparisons within and across quartiles.	We examined changes in the frequency with which topics were discussed over time. The timing results reveal that, in general, transition processes are more frequently discussed early and then teams shift to discussing action at later points during taskwork. These findings extend previous work on shifts in teamwork (e.g., Gersick 1988) and show how a team's focus changes from planning to doing over time. Encouraging teams to engage in the early planning conversations (versus jumping into action) is imperative.	The comparisons of frequencies were conducted using repeated measures analysis of variance. The analysis builds on previous comparative studies of communication (e.g., Olson et al. 1992) to look at differences among topic frequency in quartiles of data. This type of analysis demonstrates the benefit of starting with relatively simple comparisons to take stock of the many differences that exist.
Pacing – Speed of change during different parts of the process	*Data*: Social network analysis density of topic transitions. *Method*: Statistical comparisons within and across quartiles and network pictographs.	We investigated the way teams pace their conversations about team processes over time. The results provide evidence that the discussions about transition and action processes are not random. Rather teams discuss many topics at the onset of team activity and then become more streamlined as they focus on doing the work required to complete the task. These focal changes suggest leaders should allow early robust discussions but then help the team move to action.	Pacing was tested using message-to-message topic transitions to calculate the unweighted density of a directed graph. Visual representations augment the analysis by depicting the pace of team communication over time. In particular, the visualizations of density help identify shifts and plateaus in when, and how many, topics are discussed during the teams' lifecycles.
Polyphony – Aligned and overlapping events	*Data*: Patterns identified in topic communication strings. *Method*: Proportional hazards and Anderson-Gill Intensity modeling from event history analysis.	We studied the interrelationships among the timing of when teams initiate (action) or complete (transition processes) conversations. The results help us understand the impact of if and when a topic is discussed to gain insight about the way the timing of one topic being discussed aligns with or prompts another discussion to happen. Time pressured teams used a trial-and-error approach (versus the systematic approach of teams with unrestricted time) that introduced action early suggesting they may employ a bounded rational approach (Simon 1972). These insights may be useful in facilitating when teams engage in discussions about specific topics.	We used event history analysis to understand the likelihood that teams will engage in topic discussions across time given other points of interest occurred. Proportional hazards models focused on the interrelationships among the first instance of our patterns of interest. The Anderson -Gill intensity model was used to examine repetitive discussions over time. This approach may be useful to examine other processes that might include interdependent elements such as mental model convergence (McComb, 2007; Kennedy & McComb, 2010)
Sequence – The temporal ordering of events	*Data*: Topic-purpose transition matrices. *Method*: Chi-squared tests of independence, hierarchical clustering and dendrograms.	We examined the sequence of communication by looking at topic-purpose message-to-message transitions. The results confirm that teams engage in sequential discussions and provide insights into why teams may be talking about each process. For instance, teams need to engage in both divergent (i.e., idea generation) and convergent (i.e., evaluative) thinking (Paletz & Schumm, 2010); better performing teams were able to switch between these two modes more readily than lower performing teams. A team's purpose for discussing various topics changes over time and teams may need help switching how their conversations are unfolding.	After determining that the data included sequential structure using chi-squared tests of independence, we conducted hierarchical cluster analysis to identify sequential patterns that represent what and how the transition processes are discussed over time. This approach builds on the research of Olson et al. (1994) in considering patterns of change over time. Further, we also extend research from Waller (1999)and Tschan (1995, 2002) by considering how the timing of patterns evolves throughout a team's lifecycle. Dendrograms visually represent the clusters, thereby facilitating comparisons across study conditions.

rhythm, timing, pacing, polyphony, and sequence attributable to collaborative conditions and performance outcomes. More interestingly, perhaps, are the implications of these findings for how teams are managed. The original intent of the temporal dimensions was to guide planned organizational change. This notion of *planned* change may be useful for teams. Our research has shown that communication patterns exist, change over time, and have varying levels of efficacy. Given these findings, planning more purposeful team communication may be warranted through agenda setting, conversation facilitation, informal nudging, and the like.

Our most fundamental theoretical contributions, however, confirm and extend the complex view of team processes forwarded by Marks and colleagues (2001). Specifically, we link the team communication of laboratory project teams to the Marks et al. (2001) recurring phase model, which is comprised of iterative cycles of transition and action processes occurring within performance episodes. The transition processes embedded in that model are particularly helpful in understanding how teams tackle planning through mission analysis, goal specification, and strategy formulation. Our findings confirm this model by linking team communication to the action and transition processes, thereby uncovering *what* teams are doing through their conversations. Further, by tracking these processes across performance episodes, we gain perspective on *when* certain processes are most relevant and *how* the team shifts among the processes over time.

More importantly, we extend the recurring phase model in two specific ways. First, as we espoused earlier in this book, our data shows that strategy formulation may be better represented as two types of strategies, one focused on the tactical strategy (i.e., how the team will approach the task) and one about the operational strategy (i.e., who is/will do what work on the task). Separating these two strategical foci is imperative because, as our findings demonstrate, they are relevant at different points in the performance episode. This nuance was first suggested in Kennedy and McComb (2014) and is further confirmed through the research reported herein.

Our second extension of the recurring phase model is the addition of a second layer of differentiation. In addition to examining the topics being discussed, we also identified the purpose for delivering each message, which elucidated whether the team member was sending a particular message to act, agree, apply, clarify, inform, suggest, or summarize. When paired with the topic, we gained insight into both *what* teams were discussing and *why* they were discussing it. Moreover, we examined how these topic-purpose pairings clustered within quartiles for better and worse performing teams to ascertain if the sequences in which the teams engaged ultimately may help or hinder their performance. This approach advances a more granular view of the recurring phase model and offers a clearer picture of the interweaving patterns within longitudinal team communication.

Methodological Contributions

Our holistic examination of team communication would not have been feasible without the application of multiple methodologies for both analysis and visualization. To that end, one methodological contribution is the amalgamation of results from many different temporal perspectives. As the five key themes described earlier in this chapter demonstrate, we would not have been able to provide such a rich interpretation of team communication without these various perspectives. Nevertheless, our efforts employ only a fraction of the techniques available. As noted in our systematic literature review (Chap. 2), the majority of studies reviewed employed many different analysis techniques and visualization approaches. This plethora of options suggests how rich the methodological realm is for researchers interested in longitudinal, emergent phenomena. Unfortunately, the majority of papers reviewed include limited, if any, instruction regarding the analytical techniques used and the rationale behind the decisions required to appropriately apply the techniques. To address this limitation, we provided comprehensive appendices throughout this book to demonstrate how we conducted our analyses, which include specific instructions using an example dataset and providing explanations of decisions required at various steps in the process. Our approach patterned that of Pilny and Poole (2017, p. 2) in their edited volume focused on "demystifying" computational methodologies available to examine group processes. Our hope is that more researchers will provide accessible documentation of their methodological approaches, thereby encouraging more scholars to apply them in their own inquiries.

More in depth, multi-level examinations of teams are feasible with the methodologies we employed. These examinations could include, for instances, additional behaviors like physical activity, humor, gestures, artifact creation, in lieu of or in combination with team communication. Cross-level investigations into how the actions of various teams contribute to the performance of multi-team systems may be insightful. The emergence of team-level phenomena from individual-level behaviors represents another option for employing the types of methodological techniques applied herein. For example, recurrence quantification analysis may be a useful methodology to study team coordination by capturing the state of each team member at different points in time. These states could include information about movement, position, process, etc. The states of two or more members could then be assessed to visualize and examine when the members are in the same state or "in sync." This type of insight may facilitate uncovering the dynamic structure of coordination, including its complexity, predictability, and stability over time.

Study Limitations and Future Opportunities

Our study design provided the opportunity to capture, code, and examine message-by-message team communication across different temporal and media conditions. Nevertheless, limitations exist regarding the extent to which we can draw insights and generalities because of this design. While these limitations must bound the settings for our implications, they also suggest the need for future research to extend our findings into new settings and with additional features and constructs. First, we used an intellective task that has an optimal solution (McGrath 1984) for our laboratory study. In part, because the deductive design of the task was reasonably straightforward, teams did not have to overcome any ambiguity in the goals associated with the task or extensively use creative processes to construct a solution. This task, therefore, may have influenced the types of patterns in transition and action processes that we observed. Moreover, it may have minimized the need for discussing interpersonal processes such as affect, motivation, and/or conflict management. These types of processes may be more prevalent and salient when teams have to complete more creative tasks (e.g., Isen et al. 1987). As such, future researchers may want to explore different types of tasks to ascertain (1) if the patterns of communication we uncovered are generalizable to other types of tasks and (2) under what circumstances interpersonal processes may be required.

In addition to our use of one particular task we also decided to focus on transition processes. That is, we specifically examined different types of transition processes whereas we looked at action processes in the aggregate. This focus on transition processes facilitated drawing several useful insights about team communication, such as considering the nuances of strategy formulation and ultimately separating it into tactical strategy and operational strategy. An analogous investigation of the specific action processes may be useful to provide similar granular insights into how a team's work is accomplished. Such a study may help advance new research and practical implications about the way action processes are managed in terms of when, how, how much, and why action processes are discussed and/or enacted.

Another limitation may be the size and composition of our sample. We gathered data from 131 undergraduate student teams completing a laboratory task, with 120 of these teams having data at two performance episodes. The composition and size of our sample may limit the generalizability of our findings and have reduced the power of our statistical analyses, particularly when we split the data by temporal, media, and performance differences. Nevertheless, a laboratory setting mitigates exogenous factors and ensures continuity across study conditions, which together facilitated direct, meaningful comparisons. Moreover, our efforts generated multiple interesting findings about how to enhance team performance. Future researchers may find value in testing and honing our findings through additional laboratory and field studies, as well as via simulation and virtual experimentation. These studies may be particularly useful if they incorporate the plethora of communication media available, since various media foster or limit the amount, and type, of information that can be transferred. As our research findings suggest, the variability across these media may impact team members' abilities to communicate effectively. Understanding how these media influence information transfer, may uncover additional insights about longitudinal team communication.

A final limitation is not with our study, per se, but with the effort required to conduct this type of research. Our dataset took 5 years from the onset of study design to being ready for data analysis. The work included collecting, cleaning, transcribing, and coding the communicated messages, which ranged from 119 (distributed time pressured teams during the second performance episode) to 419 (in person time pressured teams during the first performance episode) messages exchanged per team. No wonder few researchers take this path, as we demonstrated in our systematic literature review (Chap. 2). At least two opportunities to diminish this limitation are worth exploring: (1) simulation and virtual experimentation and (2) surrogate measures.

A dataset comprised of coded communication strings may be useful for constructing computational models of team communication behavior that can be used to explore potential eventualities of what-if scenarios through virtual experimentation. The empirical data play a critical role in validating the simulation procedures

(Kennedy and McComb 2017). Kennedy and McComb (2014) provides an example of how data can be used to validate computational models. A portion of the data used herein (described in Chap. 3) was used to validate a computational model of team communication, thereby ensuring that model outcomes are representative of communication observed in real-world teams. This model was then employed to test various what-if scenarios regarding interventions designed to influence when teams discussed certain topics. Conducting enough experiments in a laboratory to achieve comparable results would have been cost and time prohibitive. Moreover, the results can be used to inform the design of future laboratory or field studies to ensure that the interventions with the most potential for impact, as identified through virtual experimentation, are being examined. Beyond this team-level examination of communication, the opportunities to examine longitudinal team behaviors are limitless. For instance, an expansion of Kennedy and McComb's (2014) study could focus on the input of individual team members by including information about who speaks each message. This additional information expands the set of possible what-if questions that could be explored via virtual experimentation.

Although new technologies for capturing and interpreting real-time communication are continuously being developed and improved, they may not provide the fidelity that the time-consuming task of manually coding data does. Therefore, surrogate measures may be useful. For instance, a burgeoning area of study about team neurodynamics comes from the use of electroencephalogram data. This type of data can provide greater granularity (i.e., second-by-second measurements) of brain rhythms. Researchers are using this data to examine the neurophysiologic synchronies and co-expressions of activities during dyad and group work (Delorme et al. 2002; Stevens and Galloway 2016), which may help researchers interested in constructs such as shared mental models. Other researchers are employing physiological measures to identify team-level phenomena. For example, Mønster and colleagues (2016) examined physiological synchrony among team members and identified cohesion through synchrony of smiling and conflict/tension through synchrony of skin conductance. Team researchers may benefit from employing such approaches, particularly in combination with more traditional approaches, to confirm the efficacy of the surrogate measures, as well as advance insights about team processes.

Concluding Thoughts

By examining longitudinal team communication from various temporal perspectives and via multiple methodological approaches, we are able to offer insights about the rhythm, timing, pacing, polyphony, and sequence of teams' communication exchanges. More importantly, combining these examinations within a single treatise provided us with the opportunity to offer a holistic view of team communication, which resulted in the identification of general themes representing what, how, how much, and why teams communicate as they complete their assigned task. Some of these themes apply to all teams, such as the benefit of engaging in transition processes before shifting into action. Others differed based on the conditions under which the teams were collaborating, including whether the teams were in their first or second performance episode, working with unrestricted time or under time pressure, and engaging in distributed or in person communication. Together, our findings bring to light the need to see communication, less as a random walk through team processes, and more as a planned, coordinated effort that can facilitate efficient collaboration and lead the way to effective performance.

Managerial Minutes

While we have provided suggestions for enhancing practice in each chapter, several themes cut across chapters. These themes represent specific steps team leaders and members can take to improve their team communication.

Engage in transition processes before action processes.
Engage in conversations about what you have been asked to do and how you are going to do it before you jump in and start doing it.

Recognize the changing relevance of transition processes over time.
Take time to periodically revisit what you are doing and how you are doing it to be sure the team stays on track. You also need to watch for recurring conversations that consist of clarifications without resolutions because they can be an indication that your team is struggling.

Avoid the curse of unrestricted time.
Invoke some time pressure so that your team focuses on its work and makes progress. We all seem to work more effectively as the deadline approaches.

Exploit learning over time.
Be sure that what you are learning as you complete the task is being applied to future activities. Too often teams forget about decisions they have made and then end up "recreating the wheel." But be careful, because some learned behaviors may be like bad habits and need to be adjusted.

Satisfice under time pressure.
Sometimes your team may have to move forward with incomplete information and update as they learn more. Training may be necessary to help team members devise strategies for functioning in this way.

Appendix A: Laboratory Task Documents

The documents distributed to teams at the beginning of the laboratory performance episodes are provided in this appendix. A complete task description and the procedures followed can be found in Chap. 3.

Figures A.1a, A.1b, and A.1c are the information sheets distributed to the team members. These sheets were printed on different colored paper. The text is identical except for the last paragraph. Figures A.2a and A.2b are the schedule worksheets that teams were tasked with completing. Figure A.2a was used for performance episode 1 and Fig. A.2b was used for performance episode 2.

S. McComb, D. Kennedy, *Computational Methods to Examine Team Communication*, Computational Social Sciences,
https://doi.org/10.1007/978-3-030-36159-4

a

Information Sheet Human Resources Manager

- You are to work on the problem as a team.
- You can share the information you have with your team members, but you do not have to.
- There must be a substantial agreement among your team that the problem has been solved to the best of your ability.
- You have 45 minutes to complete the task.

Best Service Store was established in 1972. It is located in the suburbs of a large metropolitan city. Best Service store retails clothing, personal accessories and personal care product for men, women and babies from established brands as well as its own brand name. The store is independent and has an established clientele and steady sales. It had not yet suffered from the competition of the nearby mall, but the manager is concerned and strives to achieve the best possible service in order to have competitive advantage. From this concern stem some of the requirements that are posed to your team.

Your team consists of a union member, production manager and human resource manager. Your team has been authorized by the management of the Best Service Store to assign the available workforce to respective time slots.

Specifically, you need to decide when (day and respective time slots) each employee will work. The schedule is constructed for a period of 7 days. You employ 10 workers that can be assigned. Each worker has a respective wage, given below. The department store is open every day from 8 a.m. till 8 p.m. The requirements for each day are stated on the Workforce Schedule Sheet as a number of employees required.

Workforce wages:

Al – 6 $/ hour	Ed – 7 $/hour	Ian – 10 $/hour
Bob – 6 $/hour	Fay – 8 $/hour	Jen – 10 $/hour
CeCe – 6 $/hour	Gail – 8 $/hour	
Dan – 7 $/hour	Hal – 9 $/hour	

As a human resources manager, you must protect your workforce by including all employees listed. In addition, no employee can work more than 10 hours/day and 50 hours/week. If you cannot follow these rules, your team will be assessed a penalty. Penalties will cost the team 1.8 times the wages of the affected employee. For example, if Al works 11 hours one day, his cost for that day will be 10*$6 for the ten hours your rule allows him to work *plus* 1*1.8*$6 for the additional hour beyond the maximum of 10 hours.

Fig. A.1 Information sheets for laboratory task. (**a**) Human resource manager information sheet, (**b**) Union representative information sheet, and (**c**) Production manager information sheet

b

Information Sheet Union Representative

- You are to work on the problem as a team.
- You can share the information you have with your team members, but you do not have to.
- There must be a substantial agreement among your team that the problem has been solved to the best of your ability.
- You have 45 minutes to complete the task.

Best Service Store was established in 1972. It is located in the suburbs of a large metropolitan city. Best Service store retails clothing, personal accessories and personal care product for men, women and babies from established brands as well as its own brand name. The store is independent and has an established clientele and steady sales. It had not yet suffered from the competition of the nearby mall, but the manager is concerned and strives to achieve the best possible service in order to have competitive advantage. From this concern stem some of the requirements that are posed to your team.

Your team consists of a union representative, production manager and human resource manager. Your team has been authorized by the management of the Best Service Store to assign the available workforce to respective time slots.

Specifically, you need to decide when (day and respective time slots) each employee will work. The schedule is constructed for a period of 7 days. You employ 10 workers that can be assigned. Each worker has a respective wage, given below. The department store is open every day from 8 a.m. till 8 p.m. The requirements for each day are stated on the Workforce Schedule Sheet as a number of employees required.

Workforce wages:

Al – 6 $/ hour	Ed – 7 $/hour	Ian – 10 $/hour
Bob – 6 $/hour	Fay – 8 $/hour	Jen – 10 $/hour
CeCe – 6 $/hour	Gail – 8 $/hour	
Dan – 7 $/hour	Hal – 9 $/hour	

As a representative of the union, you must protect your constituency by ensuring that each worker is scheduled each week and is scheduled for no less than 4 hours/day and 30 hours/week. If you cannot follow these rules, your team will be assessed a penalty. Penalties will cost the team 1.8 times the wages of the affected employee. For example, if Al works 3 hours one day, his cost for that day will be 3*$6 for the scheduled three hours *plus* 1*1.8*$6 for the additional hour that he should have been scheduled (since he must work 4 hours/day per your rule).

Fig. A.1 (continued)

c

Information Sheet Production Manager

- You are to work on the problem as a team.
- You can share the information you have with your team members, but you do not have to.
- There must be a substantial agreement among your team that the problem has been solved to the best of your ability.
- You have 45 minutes to complete the task.

Best Service Store was established in 1972. It is located in the suburbs of a large metropolitan city. Best Service store retails clothing, personal accessories and personal care product for men, women and babies from established brands as well as its own brand name. The store is independent and has an established clientele and steady sales. It had not yet suffered from the competition of the nearby mall, but the manager is concerned and strives to achieve the best possible service in order to have competitive advantage. From this concern stem some of the requirements that are posed to your team.

Your team consists of a union member, production manager and human resource manager. Your team has been authorized by the management of the Best Service Store to assign the available workforce to respective time slots.

Specifically, you need to decide when (day and respective time slots) each employee will work. The schedule is constructed for a period of 7 days. You employ 10 workers that can be assigned. Each worker has a respective wage, given below. The department store is open every day from 8 a.m. till 8 p.m. The requirements for each day are stated on the Workforce Schedule Sheet as a number of employees.

Workforce wages:

Al – 6 $/ hour	Ed – 7 $/hour	Ian – 10 $/hour
Bob – 6 $/hour	Fay – 8 $/hour	Jen – 10 $/hour
CeCe – 6 $/hour	Gail – 8 $/hour	
Dan – 7 $/hour	Hal – 9 $/hour	

As a production manager, you must ensure that your team achieves the schedule with the lowest possible cost. In addition, the daily requirements need to be satisfied exactly. If you cannot follow these rules, your team will be assessed a penalty. Penalties will cost the team 1.8 times the wages of the most expensive employee. For example, if your schedule has too many workers, the most expensive worker will be charged at a rate of 1.8 times his/her hourly wage for all the hours s/he is not needed. So, if the schedule calls for two workers all day and during one hour of your schedule Al, Bob, and Dan are assigned to work, your team will incur a penalty of 1*1.8*$7 for that hour.

Fig. A.1 (continued)

a

WORKFORCE SCHEDULE WORKSHEET SCENARIO 1						Team _____
Hour	Day 1	Day 2	Day 3	Day 4	Day 5	Day 6	Day 7
8 – 10 am							
10 – 12 am							
12 – 2 pm							
2 – 4 pm							
4 – 6 pm							
6 – 8 pm							
Employees required per shift	**3**	**3**	**5**	**4**	**6**	**7**	**4**

b

WORKFORCE SCHEDULE WORKSHEET SCENARIO 2						Team _____
Hour	Day 1	Day 2	Day 3	Day 4	Day 5	Day 6	Day 7
8 – 10 am	2	2	3	2	4	4	4
10 – 12 am	2	2	3	2	4	4	4
12 – 2 pm	3	3	4	3	4	5	4
2 – 4 pm	3	3	4	3	4	5	4
4 – 6 pm	3	3	5	4	6	7	5
6 – 8 pm	3	3	5	4	6	7	5

Fig. A.2 Laboratory task worksheets. (**a**) Performance episode 1 worksheet, and (**b**) Performance episode 2 worksheet

References

Allison, P. D. (1984). *Event history analysis: Regression for longitudinal event data*. Newbury Park: Sage Publications.
Allison, P. D. (1995). *Survival analysis using SAS: A practical guide*. Cary: SAS Institute.
Ancona, D., & Chong, C. (1999). Cycles and synchrony: The temporal role of context in team behavior. In R. Wageman (Ed.), *Research on managing groups and teams* (Vol. 2, pp. 33–48). Greenwich: JAI Press.
Anderson, P. K., & Gill, R. D. (1982). Cox's regression model for counting processes: A large sample study. *The Annals of Statistics, 10*(4), 1100–1120.
Andres, H. P. (2013). Team cognition using collaborative technology: A behavioral analysis. *Journal of Managerial Psychology, 28*(1), 38–54.
Aufegger, L., Bicknell, C., Soane, E., Ashrafian, H., & Darzi, A. (2019). Understanding health management and safety decisions using signal processing and machine learning. *BMC Medical Research Methodology, 19*, 121. https://doi.org/10.1186/s12874-019-0756-2.
Bachrach, D. G., Bendoly, E., & Podsakoff, P. M. (2001). Attributions of the "causes" of group performance as an alternative explanation of the relationship between organizational citizenship behavior and organizational performance. *Journal of Applied Psychology, 86*, 1285–1295.
Bales, R. F. (1950). *Interaction process analysis: A method for the study of small groups*. Reading: Addison-Wesley.
Bales, R. F., & Strodtbeck, F. L. (1951). Phases in group problem-solving. *Journal of Abnormal and Social Psychology, 46*, 485–495.
Ballard, D. I., Tschan, F., & Waller, M. J. (2008). All in the timing: Considering time at multiple stages of group research. *Small Group Research, 39*(3), 328–351.
Barth, S., Schraagen, J. M., & Schmettow, M. (2015a). Network measures for characterizing team adaptation processes. *Ergonomics, 58*(8), 1287–1302.
Barth, S., Schraagen, J. M., & Schmettow, M. (2015b). Network measure for characterizing team adaptation processes. *Ergonomics, 58*(8), 1287–1302.
Bartunek, J. M., & Woodman, R. W. (2015). Beyond Lewin: Toward a temporal approximation of organization development and change. *Annual Review of Organizational Psychology and Organizational Behavior, 2*, 157–182.
Benzer, J. K., Mohr, D. C., Evans, L., Young, G., Meterko, M. M., Moore, S. C., Seibert, M. J., Osatuke, K., Stolzmann, K. L., White, B., & Charns, M. P. (2016). Team process variation across diabetes quality care trajectories. *Medical Care Research and Review, 73*(5), 565–589.
Bergan, J. R., & Tombari, M. L. (1975). The analysis of verbal interactions occurring during consultation. *Journal of School Psychology, 13*, 309–318.

S. McComb, D. Kennedy, *Computational Methods to Examine Team Communication*, Computational Social Sciences,
https://doi.org/10.1007/978-3-030-36159-4

Berry, G. R. (2006). Can computer-mediated asynchronous communication improve team processes and decision making? *Journal of Business Communication, 43*(4), 344.366.

Bohm, A. (2004). Theoretical coding: Text analysis in grounded theory. In U. Flick, E. von Kardorff, & I. Steinki (Eds.), *A companion to qualitative research* (pp. 270–275). London: Sage.

Borgatti, S. P. (2002). *NetDraw network visualization*. Harvard: Analytic Technologies.

Borgatti, S. P., Everett, M. G., & Freeman, L. C. (2002). *UCINET 6 for windows: Software for social network analysis*. Harvard: Analytic Technologies.

Borgatti, S. P., Everett, M. G., & Johnson, J. C. (2018). *Analyzing social networks* (2nd ed.). Los Angeles: Sage.

Bourbousson, J., R'Kiouak, M., & Eccles, D. W. (2015). The dynamics of team coordination: A social network analysis as a window to shared awareness. *European Journal of Work and Organizational Psychology, 24*(5), 742–760.

Bowers, C. A., Jentsch, F., Salas, E., & Braun, C. C. (1998). Analyzing communication sequences for team training needs assessment. *Human Factors, 40*(4), 672–679.

Brooks-Buza, H., Fernandez, R., & Stenger, J. P. (2011). The use of in situ simulation to evaluate teamwork and system organization during a pediatric dental clinic emergency. *Simulation in Healthcare-Journal of the Society for Simulation in Healthcare, 6*(2), 101–108.

Brown, A., Dickinson, H., & Kelaher, M. (2018). Governing the quality and safety of healthcare: A conceptual framework. *Social Science and Medicine, 202*, 99–107.

Bunger, A. C., & Lengnicki-Hall, R. (2018). Do learning collaborative strengthen communication? A comparison of organizational team communication networks over time. *Health Care Management Review, 43*(1), 50–60.

Calder, L. A., Mastoras, G., Rahimpour, M., Sohmer, B., Weitzman, B., Cwinn, A. A., Hobin, T., & Parush, A. (2017). Team communication patterns in emergency resuscitation: A mixed methods qualitative analysis. *International Journal of Emergency Medicine, 10*, 24.

Caldwell, B. S., & Everhart, N. C. (1998). Information flow and development of coordination in distributed supervisory control teams. *International Journal of Human-Computer Interaction, 10*(1), 51–70.

Cannon-Bowers, J., Salas, E., & Converse, S. (1993). Shared mental models in expert team decision making. In N. Castellan Jr. (Ed.), *Individual and group decision making* (pp. 221–246). Hillsdale: Lawrence Erlbaum.

Chi, M. T. H. (1997). Quantifying qualitative analyses of verbal data: A practical guide. *The Journal of Learning Sciences, 6*(3), 271–315.

Chiocchio, F. (2007). Porject team performance: A study of electronic task and communication coordination. *Project Management Journal, 38*(1), 97–109.

Chiu, M. M., & Lehmann-Willenbrock, N. (2016). Statistical discourse analysis: Modeling sequences of individual actions during group interactions across time. *Group Dynamics, 20*(3), 242–258.

Cohen, S., & Bailey, D. (1997). What makes teams work: Group effectiveness research from the shop floor to the executive suite. *Journal of Management, 23*, 239–290.

Colquitt, J. A., Hollenbeck, J. R., Ilgen, D. R., LePine, J. A., & Sheppard, L. (2002). Computer-assisted communication and team decision-making performance: The moderating effect of openness to experience. *Journal of Applied Psychology, 87*(2), 402–410.

Cooke, N. J., Gorman, J. C., Duran, J. L., & Taylor, A. R. (2007). Team cognition in experienced command-and-control teams. *Journal of Experimental Psychology: Applied, 13*(3), 146–157.

Cronin, M. A., Weingart, L. R., & Todorova, G. (2011). Dynamics in groups: Are we there yet? *The Academy of Management Annals, 5*, 571–612.

Daft, R. L., & Lengel, R. H. (1986). Organizational information requirements, media richness and structural design. *Management Science, 32*(5), 554–571.

David, L. Z., & Schraagen, J. M. (2018). Analysing communication dynamics at the transaction level: The case of Air France flight 447. *Cognition, Technology and Work, 20*, 637–649.

Davis, W. A., Jones, S., Crowell-Kuhnberg, A. M., O'Keeffe, D., Boyle, K. M., Klainer, S. B., Smink, D. S., & Yule, S. (2016). Operative team communication during simulated emergencies: Too busy to respond? *Surgery, 161*(5), 1348–1356.

DeChurch, L. A., & Haas, C. D. (2008). Examining team planning through an episodic lens: Effects of deliberate, contingency, and reactive planning on team effectiveness. *Small Group Research, 39*(5), 542–568.

Delorme, A., Rousselet, G., Mace, M., & Fabre-Thorpe, M. (2002). Interaction of bottom-up and top-down processing in the fast visual analysis of natural scenes. *Cognitive Brain Research, 19*, 103–113.

DeMatteo, J. S., Eby, L. T., & Sundsrom, E. (1998). Team-based rewards: Current empirical evidence and directions for future research. *Research in Organizational Behavior, 20*, 141–183.

Diaz Granados, D., Dow, A. W., Appelbaum, N., Mazmanian, P. E., & Retchin, S. M. (2018). Interprofessional practice in different patient care settings: A qualitative exploration. *Journal of Interprofessional Care, 32*(2), 151–159.

Driskell, J. E., Salas, E., & Johnston, J. H. (1999). Does stress lead to a loss of team perspective? *Group Dynamics, 3*, 1–12.

Dunn, J. C., Lewandowsky, S., & Kirsner, K. (2002). Dynamics of communication in emergency management. *Applied Cognitive Psychology, 16*, 719–737.

Earley, C. (1994). Self or group? Cultural effects of training on self-efficacy and performance. *Administrative Science Quarterly, 39*, 89–108.

Eckmann, J. P., Kamphorst, S. O., & Ruelle, D. (1987). Recurrence plots of dynamical systems. *Europhysics Letters, 4*, 973–977.

Ekker, K. (2016). Emergency management training: Handling rich qualitative and quantitative data. *Journal of Intelligent and Fuzzy Systems, 31*, 939–948.

Ellis, A. P. J. (2006). System breakdown: The role of mental models and transactive memory in the relationship between acute stress and team performance. *Academy of Management Journal, 49*(3), 576–589.

Entin, E. E., Diedrich, F. J., & Rubineau, B. (2003). Adaptive communication patterns in different organizational structures, Proceedings of the human factors and ergonomics society 47th annual meeting, pp. 405–409.

Erez, M. (1977). Feedback: A necessary condition for the goal setting-performance relationship. *Journal of Applied Psychology, 62*, 624–627.

Ericksen, J., & Dyer, L. (2004). Right from the start: Exploring the effects of early team events on subsequent project team development and performance. *Administrative Science Quarterly, 49*, 438–471.

Espinosa, J. A., Nan, N., & Carmel, E. (2015). Temporal distance, communication patterns, and task performance in teams. *Journal of Management Information Systems, 32*(1), 151–191.

Fernandez, R., Pearce, M., Grand, J. A., Rench, T. A., Jones, K. A., Chao, G. T., & Kozlowski, S. W. J. (2013). Evaluation of a computer-based educational intervention to improve medical teamwork and performance during simulated patient resuscitations. *Critical Care Medicine, 41*(11), 2551–2562.

Fiore, S. M., Smith-Jentsch, K. A., Salas, E., Warner, N., & Letsky, M. (2010a). Towards an understanding of macrocognition in teams: Developing and defining complex collaborative processes and products. *Theoretical Issues in Ergonomics Science, 11*(4), 250–271.

Fiore, S. M., Rosen, M. A., Smith-Jentsch, K. A., Salas, E., Letsky, M., & Warner, N. (2010b). Toward an understanding of mactrocognition in teams: Predicting processes in complex collaborative contexts. *Human Factors, 52*(2), 203–224.

Fleştea, A. M., Fodor, O. C., Curşeu, P. L., & Miclea, M. (2017). "We didn't know anything, it was a mess!" Emergent structures and the effectiveness of a rescue operation multi-team system. *Ergonomics, 60*(1), 44–58.

Fusaroli, R., Bjørndahl, J. S., Roepstorff, A., & Tylén, K. (2016). A heart for interaction: Shared physiological dynamics and behavioral coordination in a collective, creative construction task. *Journal of Experimental Psychology: Human Perception and Performance, 42*(9), 1297–1310.

Futoran, G. C., Kelly, J. R., & McGrath, J. E. (1989). TEMPO: A time-based system for analysis of group interaction process. *Basic and Applied Social Psychology, 10*, 211–232.

Gerpott, F. H., Lehmann-Willenbrock, N., Voelpel, S. C., & van Vugt, M. (2019). It's not just what is said, but when it's said: A temporal account of verbal behaviors and emergent leadership in self-managed teams. *Academy of Management Journal, 62*(3), 717–738.

Gersick, C. J. G. (1988). Time and transition in work teams: Toward a new model of group development. *Academy of Management Journal, 31*(1), 9–41.

Gersick, C. J. G. (1989). Marking time: Predictable transitions in task groups. *Academy of Management Journal, 32*(2), 274–309.

Gersick, C. (1991). Revolutionary change theories: A multilevel exploration of the punctuated equilibrium paradigm. *Academy of Management Review, 16*, 10–36.

Gittinger, M., Brolliar, S. M., Grand, J. A., Nichol, G., & Fernandez, R. (2017). Using simulation as an investigational methodology to explore the impact of technology on team communication and patient management: A pilot evaluation of the effect of an automated compression device. *Simulation in Healthcare, 12*(3), 139–147.

Gladstein, D. L. (1984). Groups in context: A model of task group effectiveness. *Administrative Science Quarterly, 29*, 499–517.

Goodman, R. A., & Goodman, L. P. (1976). Some management issues in temporary systems: A study of professional development and manpower-the theater case. *Administrative Science Quarterly, 21*(3), 494–501.

Gorman, J. C., & Cooke, N. J. (2011). Changes in team cognition after a retention interval: The benefits of mixing it up. *Journal of Experimental Psychology: Applied, 17*(4), 303–319.

Gorman, J. C., Amazeen, P. G., & Cooke, N. J. (2010). Team coordination dynamics. *Nonlinear Dynamics, Psychology, and Life Sciences, 14*(3), 265–289.

Gorman, J. C., Cooke, N. J., Amazeen, P. G., & Fouse, S. (2012a). Measuring patterns in team interaction sequences using a discrete recurrence approach. *Human Factors, 54*(4), 503–517.

Gorman, J. C., Hessler, E. E., Amazeen, P. G., Cooke, N. J., & Shope, S. M. (2012b). Dynamical analysis in real time: Detecting perturbations to team communication. *Ergonomics, 55*(8), 825–839.

Gorman, J. C., Martin, M. J., Dunbar, T. A., Stevents, R. H., Galloway, T. L., Amazeen, P. G., & Likens, A. D. (2016). Cross-level effects between neurophysiology and communication during team training. *Human Factors, 58*(1), 181–199.

Grand, J. A., Braun, M. T., Kuljanin, G., Kozlowski, S. W. J., & Chao, G. T. (2016). The dynamics of team cognition: A process-oriented theory of knowledge emergence in teams. *Journal of Applied Psychology, 101*(10), 1353–1385.

Griffin, A., & Hauser, J. R. (1992). Patterns of communication among marketing, engineering and manufacturing – A comparison between two new product teams. *Management Science, 38*(3), 360–373.

Gundrosen, S., Andenæs, E., Aadahl, P., & Thomassen, G. (2016). Team talk and team acivity in simulated medical emergencies: A discourse analytical approach. *Scandinavian Journal of Trauma, Resuscitation and Emergency Medicine, 24*, 135. https://doi.org/10.1186/s13049-016-0325-1.

Hackman, J. R., & Morris, C. G. (1975). Group tasks, group interaction processes, and group performance effectiveness: A review and prposed integration. In L. Berkowitz (Ed.), *Advances in experimental social psychology* (Vol. 8, pp. 45–99). New York: Academic.

Hackman, J., & Wageman, R. (2005). A theory of coaching. *Academy of Management Review, 30*, 269–287.

Hackman, J., Brousseau, K. R., & Weiss, J. A. (1976). The interaction of task design and group performance strategies in determining group effectiveness. *Organizational Behavior and Human Performance, 16*, 350–365.

Hammond, J. M., Harvey, C. M., Koubek, R. J., Compton, W. D., & Darisipudi, A. (2005). Distributed collaborative design teams: Media effects on design processes. *International Journal of Human-Computer Interaction, 18*(2), 145–165.

Havakhor, T., & Sabherwal, R. (2018). Team processes in virtual knowledge teams: The effects of reputation signals and network density. *Journal of Management Information Systems, 35*(1), 266–318.

Hewes, D. E., & Poole, M. S. (2012). The analysis of group interaction processes. In A. B. Hollingshead & M. S. Poole (Eds.), *Research methods for studying groups and teams* (pp. 329–357). New York: Routledge.

Hirst, G. (2009). Effects of membership change on open discussion and team performance: The moderating role of team tenure. *European Journal of Work and Organizational Psychology, 18*(2), 231–249.

Hoogeboom, M. A. M. G., & Wilderom, C. P. M. (2015). Effect leader behaviors in regularly held staff meetings: Surveyed vs. videotaped and video-coded observations. In J. A. Allen, N. Lehmann-Willenbrock, & S. G. Rogelberg (Eds.), *The Cambridge handbook of meeting science* (pp. 381–412). Cambridge: Cambridge University Press.

Hoogeboom, M. A. M. G., & Wilderom, C. P. M. (2019). A complex adaptive systems approach to real-life team interaction patterns, task context, information sharing, and effectiveness. *Group and Organization Management*,. online first. https://doi.org/10.1177/1059601119854927.

Hua, Y., Becker, F., Wurmser, T., Bliss-Holtz, J., & Hedges, C. (2012). Nursing team communication patterns, quality of care, and patient safety. *Health Environments Research and Design Journal, 6*(1), 8–38.

Ilgen, D. R., Hollenbeck, J. R., Johnson, M., & Jundt, D. (2005). Teams in organizations: From input-process-output models to IMOI models. *Annual Review of Psychology, 56*, 517–543.

Isen, A. M., Daubman, K. A., & Nowicki, G. P. (1987). Positive affect facilitates creative problem solving. *Journal of Personality and Social Psychology, 52*(6), 1122–1131.

Janicik, G. A., & Bartel, C. A. (2003). Talking about time: Effects of temporal planning and time awareness norms on group coordination and performance. *Group Dynamics Theory, Research, and Practice, 7*, 122–134.

Kalia, A. K., Buchler, N., DeCostanza, A., & Singh, M. P. (2017). Computing team process measures from the structure and content of broadcast collaborative communication. *IEEE Transaction on Computational Social Systems, 4*(2), 26–39.

Kanawattanachi, P., & Yoo, Y. (2007). The impact of knowledge coordination on virtual team performance over time. *MIS Quarterly, 31*(4), 783–808.

Katzenbach, J., & Smith, D. (1999). *The wisdom of teams: Creating the high-performance organization*. Cambridge, MA: Harvard Business School Press.

Kauffeld, S. (2006). *Measuring, evaluating, and developing competencies—A process analytical approach for groups*. Stuttgart: Schäffer-Poeschel.

Kauffeld, S., & Lehmann-Willenbrock, N. (2012). Meetings matter: Effects of team meetings on team and organizational success. *Small Group Research, 43*(2), 130–158.

Kauffeld, S., & Meyers, R. (2009). Complaint and solution-oriented circles: Interaction patterns in work group discussion. *European Journal of Work and Organizational Psychology, 18*(3), 267–294.

Kelly, J. R., & McGrath, J. E. (1985). Effects of time limits and task types on task performance and interaction of four-person groups. *Journal of Personality and Social Psychology, 49*(2), 395–407.

Kennedy, D. M., & McComb, S. A. (2010). Examining the mental model convergence process through team communication. *Theoretical Issues of Ergonomic Science, 11*(4), 340–358.

Kennedy, D. M., & McComb, S. A. (2014). When teams shift among processes. *Journal of Applied Psychology, 99*(5), 784–815.

Kennedy, D. M., & McComb, S. A. (2017). Simulation and virtual experimentation: Grounding with empirical data. In A. Pilny & M. S. Poole (Eds.), *Group processes: Computational and data driven approaches*. New York: Springer.

Klein, G. (2001). Features of team coordination. In M. McNeese, E. Salas, & M. Endsley (Eds.), *New trends in cooperative activities: Understanding system dynamics in complex environments* (pp. 68–95). Santa Monica: Human Factors and Ergonomics Society.

Klein, G., Phillips, J. K., Rall, E. L., & Peluso, D. A. (2007). A data-frame theory of sensemaking. In R. T. Hoffman (Ed.), *Expertise out of context: Proceedings of the 6th international conference on naturalistic decision making* (pp. 113–156). New York: Psychology Press.

Klein, G., Wiggins, S., & Dominguez, C. O. (2010). Team sensemaking. *Theoretical Issues in Ergonomics Science, 11*(4), 304–320.

Klonek, F. E., Quera, V., Burba, M., & Kauffeld, S. (2016). Group interactions and time: Using sequential analysis to study group dynamics in project meetings. *Group Dynamics, 20*(3), 209–222.

Knight, A. P., Kennedy, D. M., & McComb, S. A. (2016). Using recurrence analysis to examine group dynamics. *Group Dynamics, Theory, Research, and Practice, 20*(3), 223–241.

Kolbe, M., Grote, G., Waller, M. J., Wacker, J., Grande, B., Burtscher, M. J., & Spahn, D. R. (2014a). Monitoring and talking to the room: Autochthonous coordination patterns in team interaction and performance. *Journal of Applied Psychology, 99*(6), 1254–1267.

Kolbe, M., Grote, G., Waller, M. J., Wacker, J., Grande, B., & Burtscher, M. J. (2014b). Monitoring and talking to the room: Autochthonous coordination patterns in team interaction and performance. *Journal of Applied Psychology, 99*(6), 1254–1267.

Kozlowski, S. W. (2015). Advancing research on team process dynamics theoretical, methodological, and measurement considerations. *Organizational Psychology Review, 5*(4), 270–299.

Kozlowski, S. W. J., Gully, S. M., Nason, E., & Smith, E. (1999). Developing adaptive teams: A theory of compilation and performance across levels and time. In D. Ilgen & E. Pulakos (Eds.), *The changing nature of performance, implications for staffing, motivation, and development* (pp. 240–292). San Francisco: Jossey-Bass.

Kozlowski, S. W. J., Chao, G. T., Grand, J. A., Braun, M. T., & Kuljanin, G. (2013). Advancing multilevel research design: Capturing the dynamics of emergence. *Organizational Research Methods, 16*, 581–615.

Kozlowski, S. W. J., Chao, G. T., Chang, C.-H., & Fernandez, R. (2016). Team dynamics: Using "big data" to advance the science of team effectiveness. In S. Tonidandel, E. King, & J. Cortina (Eds.), *Big data at work: The data science revolution and organizational psychology* (pp. 272–309). New York: Routledge Academic.

Lee, S. W., & Jamison, T. R. (2003). Including the FBA process in student assistance teams: An exploratory study of team communications and intervention selection. *Journal of Educational and Psychological Consultation, 14*(2), 209–239.

Leenders, R. T. A. J., Contractor, N. S., & DeChurch, L. A. (2016). Once upon a time: Understanding team processes as relational event networks. *Organizational Psychology Review, 6*(1), 92–115.

Lehmann-Willenbrock, N., & Allen, J. A. (2014). How fun are your meetings? Investigating the relationship between humor patterns in team interactions and team performance. *Journal of Applied Psychology, 99*(6), 1278–1287.

Lehmann-Willenbrock, N., & Allen, J. A. (2017). Well, now what do we do? Wait ... : A group process analysis of meeting lateness. *International Journal of Business Communication.* https://doi.org/10.1177/2329488417696725.

Lehmann-Willenbrock, N., & Allen, J. A. (2018). Modeling temporal interaction dynamics in organizational settings. *Journal of Business and Psychology, 33*, 325–344.

Lehmann-Willenbrock, N., & Chiu, M. M. (2017). Igniting and resolving content disagreements during team interactions: A statistical discourse analysis of team dynamics at work. *Journal of Organizational Behavior, 39*, 1142–1162.

Lehmann-Willenbrock, N., Meyers, R. A., Kauffeld, S., Neininger, A., & Henschel, A. (2011). Verbal interaction sequences and group mood: Exploring the role of team planning communication. *Small Group Research, 42*(6), 639–668.

Lehmann-Willenbrock, N., Allen, J. A., & Kauffeld, S. (2013). A sequential analysis of procedural meeting communication: How teams facilitate their meetings. *Journal of Applied Communication Research, 41*(4), 365–388.

Lehmann-Willenbrock, N., Allen, J. A., & Meinecke, A. L. (2014). Observing culture: Differences in U.S.-American and German team meeting behaviors. *Group Processes and Intergroup Relations, 17*(2), 252–271.

Lehmann-Willenbrock, N., Meinecke, A. L., Rowold, J., & Kauffeld, S. (2015). How transformational leadership works during team interactions: A behavioral process analysis. *The Leadership Quarterly, 26*, 1017–1033.

Lehmann-Willenbrock, N., Beck, S. J., & Kauffeld, S. (2016). Emergent team roles in organizational meetings: Identifying communication patterns via cluster analysis. *Communication Studies, 67*(1), 37–57.

Lehmann-Willenbrock, N., Chiu, M. M., Lei, Z., & Kauffeld, S. (2017). Understanding positivity within dynamic team interactions: A statistical discourse analysis. *Group and Organization Management, 42*(1), 39–78.

Lepine, J. A., Piccolo, R. F., Jackson, C. L., Mathieu, J. E., & Saul, J. R. (2008). A meta-analysis of teamwork processes: Tests of a multidimensional model and relationships with team effectiveness criteria. *Personnel Psychology, 61*, 273–307.

Lewin, K. (1947). Frontiers in group dynamics: Concept, method and reality in social science; social equilibrium and social change. *Human Relations, 1*(1), 5–41.

Lingard, L., Reznick, R., DeVito, I., & Espin, S. (2002a). Forming professional identities on the health care team: Discursive constructions of the 'other' in the operating room. *Medical Education, 36*, 728–734.

Lingard, L., Reznick, R., Espin, S., Regehr, G., & DeVito, I. (2002b). Team communications in the operating room: Talk patterns, sites of tension, and implications for novices. *Academic Medicine, 77*(3), 232–237.

Littlepage, G. E., Hein, M. B., Moffett, R. G., III, Craig, P. A., & Georgiou, A. M. (2016). Team training for dynamic cross-functional teams in aviation: Behavioral, cognitive, and performance outcomes. *Human Factors, 58*(8), 1275–1288.

Mancuso, M. P., Dziadkowiec, O., Kleiner, C., Halverson-Carpenter, K., Link, T., & Barry, J. (2016). Crew resource management for obstetric and neonatal teams to improve communication during cesarean births. *Journal of Obstetric, Gynecologic and Neonatal Nursing, 45*(4), 502–514.

Marks, M. A., Mathieu, J. E., & Zaccaro, S. J. (2001). A temporally based framework and taxonomy of team processes. *Academy of Management Review, 26*(3), 356–376.

Marks, M. A., DeChurch, L. A., Mathieu, J. E., Panzer, F. J., & Alonso, A. (2005). Teamwork in multiteam systems. *Journal of Applied Psychology, 90*(5), 964–971.

Martínez Ayala, D. F., Balakumar, B., McComb, S. A., & Pattipati, K. R. (2017). Markov modeling and analysis of team communication. *IEEE Transactions on Systems, Man and Cybernetics: Systems*. https://doi.org/10.1109/TSMC.2017.2748985.

Martínez-Moreno, E., Zornoza, A., González-Navarro, P., & Thompson, L. F. (2012). Investigating face-to-face and virtual teamwork over time: When does early task conflict trigger relationship conflict? *Group Dynamics, 16*(3), 159–171.

Marwan, N. (2008). A historical review of recurrence plots. *The European Physical Journal Special Topics, 164*(1), 3–12.

Marwan, N., Carmen Romano, M., Thiel, M., & Kurths, J. (2007). Recurrence plots for the analysis of complex systems. *Physics Reports, 438*, 237–329.

Mathieu, J. E., & Rapp, T. L. (2009). Laying the foundation for successful team performance trajectories: The roles of team charters and performance strategies. *Journal of Applied Psychology, 94*(1), 90–103.

Mathieu, J. E., Heffner, T. S., Goodwin, G. F., Salas, E., & Cannon-Bowers, J. A. (2000). The influence of shared mental models on team process and performance. *Journal of Applied Psychology, 85*(2), 273–283.

Mathieu, J. E., Heffner, T. S., Goodwin, G. F., Cannon-Bowers, J. A., & Salas, E. (2005). Scaling the quality of teammates' mental models: Equifinality and normative comparisons. *Journal of Organizational Behavior, 26*, 37–56.

Mathieu, J., Maynard, M. T., Rapp, T., & Gilson, L. (2008). Team effectiveness 1997-2007: A review of recent advancements and a glimpse into the future. *Journal of Management, 34*(3), 410–476.

McComb, S. (2007). Mental model convergence: The shift from being an individual to being a team member. In F. Dansereau & F. J. Yammarino (Eds.), *Multi-level issues in organizations and time* (Vol. 6, pp. 95–147). Oxford: Elsevier Science Ltd..

McComb, S., Kennedy, D., Perryman, R., Warner, N., & Letsky, M. (2010). Temporal patterns of mental model convergence: Implications for distributed teams interacting in electronic collaboration spaces. *Human Factors, 52*(2), 264–281.

McComb, S. A., Kennedy, D. M. & Perryman, R. (2014). Identifying structures in team communication using hierarchical clustering, *Proceedings of the 2014 ISE Research Conference*.

McEwan, D., Zumbo, B. D., Eys, M. A., & Beauchamp, M. R. (2018). The development and psychometric properties of the multidimensional assessment of teamwork in sport. *Journal of Sport and Exercise Psychology, 40*, 60–72.

McGrath, J. E. (1984). *Groups: Interaction and performance*. Englewood Cliffs: Prentice Hall.

McGrath, J. E., & Altman, I. (1966). *Small group research: A synthesis and critique of the field*. New York: Holt, Rinehart and Winston.

McKendrick, R., Shaw, T., de Visser, E., Saqer, H., Kidwell, B., & Parasuraman, R. (2014). Team performance in networked supervisory control of unmanned air vehicles: Effects of automation, working memory, and communication content. *Human Factors, 56*(3), 463–475.

Mehta, A., Feild, H., Armenakis, A., & Mehta, N. (2009). Team goal orientation and team performance: The mediating role of team planning. *Journal of Management, 35*, 1026–1046.

Meinecke, A. L., de Sanchez, C. S. H., Lehmann-Willenbrock, N., & Buengeler, C. (2019). Using state space grids for modeling temporal team dynamics. *Frontiers in Psychology, 10*, 863. https://doi.org/10.3389/fpsyg.2019.00863.

Meyers, R. A., & Seibold, D. R. (2012). Coding group interaction. In A. B. Hollingshead & M. S. Poole (Eds.), *Research methods for studying groups and teams* (pp. 329–357). New York: Routledge.

Meyers, R. A., Seibold, D. R., & Brashers, D. E. (1991). Argument in initial group decision making discussions: Refinement of a coding scheme and a descriptive quantitative analysis. *Western Journal of Speech Communication, 55*, 47–68.

Michelson, K., Clayman, M. L., Ryan, C., Emanuel, L., & Frader, J. (2017). Communication during pediatric intensive care unit family conferences: A pilot study of content, communication and parent perceptions. *Health Communication, 32*(10), 1225–1232.

Miller, A. (2004). A work domain analysis framework for modeling intensive care unit patients. *Cognition Technology and Work, 6*, 207–222.

Miller, A., Scheinkestel, C., & Joseph, M. (2009). Coordination and continuity of intensive care unit patient care. *Human Factors, 51*(3), 354–367.

Miller, A., Weinger, M. B., Buerhaus, P., & Dietrich, M. S. (2010). Care coordination in intensive care units: Communicating across information spaces. *Human Factors, 52*(2), 147–161.

Mitchell, T. R., & James, L. R. (2001). Building better theory: Time and the specification of when things happen. *Academy of Management Review, 26*(4), 530–547.

Mønster, D., Håkonsson, D. D., Eskildsen, J. K., & Wallot, S. (2016). Physiological evidence of interpersonal dynamics in a cooperative production task. *Physiology and Behavior, 156*, 24–34.

Morales, N. A., & Martin, A. (2018). Evaluating a stakeholder-driven bass regulation change in Florida: Agency staff and stakeholder perspectives. *Fisheries, 43*(8), 361–367.

Mueller, N. E., Kennedy, D. M., & McComb, S. A., (2019). Data visualizations depicting longitudinal team communication. Purdue University Research Repository. https://doi.org/10.4231/2FRC-W869.

Mundt, M. P., & Zakletskaia, L. I. (2018). Putting the pieces together: EHR communication and diabetes patient outcomes. *The American Journal of Managed Care, 24*(10), 462–468.

Olekalns, M., Brett, J. M., & Weingart, L. R. (2003). Phases, transitions and interruptions: Modeling processes in multi-party negotiations. *The International Journal of Conflict Management, 14*(3/4), 191–211.

Olson, G., Olson, J., Carter, M., & Storrosten, M. (1992). Small group design meetings: An analysis of collaboration. *Human-Computer Interaction, 7*(4), 347–374.

Olson, G. M., Herbsleb, J. D., & Rueter, H. (1994). Characterizing the sequential structure of interactive behaviors through statistical and grammatical techniques. *Human-Computer Interaction, 9*(4), 427–472.

Paletz, S. B. F., & Schummn, C. D. (2010). A social-cognitive framework of multidisciplinary team innovation. *Topics in Cognitive Science, 2*, 73–95.

Pandhi, N., Kraft, S., Berkson, S., Davis, S., Kamnetz, S., Koslov, S., Trowbridge, E., & Caplan, W. (2018). Developing primary care teams prepared to improve quality: A mixed-methods evaluation and lessons learned from implementing a microsystems approach. *BMC Health Services Research, 18*, 847.

Parush, A., Kramer, C., Foster-Hunt, T., McMullan, A., & Momtahan, K. (2014). Exploring similarities and differences in teamwork across diverse healthcare contexts using communication analysis. *Cognition, Technology and Work, 16*, 47–57.

Patrashkova, R. R., & McComb, S. A. (2004). Exploring why more communication is not better: Insights from a computational model of cross-functional teams. *Journal of Engineering and Technology Management, 21*, 83–114.

Pavitt, C. (1992). Describing know-how about group discussion procedure: Must the representation be recursive? *Communication Studies, 43*(3), 150–170.

Pavitt, C., & Johnson, K. K. (2001). The association between group procedural MOPs and group discussion procedure. *Small Group Research, 32*(5), 595–624.

Pilny, A., & Poole, M. S. (2017). *Group processes: Computational and data driven approaches*. New York: Springer.

Pilny, A., Schecter, A., Poole, M. S., & Contractor, N. (2016). An illustration of the relational event model to analyze group interaction processes. *Group Dynamics: Theory, Research, and Practice, 20*(3), 181–195.

Poole, M. S. (1983). Decision development in small groups II: A study of multiple sequences in decision making. *Communication Monographs, 50*(3), 206–232.

Pridmore, J., & Phillips-Wren, G. (2011). Assessing decision making quality in face-to-face teams *versus* virtual teams in a virtual world. *Journal of Decision Systems, 20*(3), 283–308.

Priest, H. A., Guthrie, J., Burke, C. S., Bowers, C., & Salas, E. (2004). Are there differences in teamwork across team type? Proceedings of the human factors and ergonomics society 48th annual meeting, pp. 2582–2586.

Pruitt, D. G., & Carnevale, P. J. (1982). The development of integrative agreements in social conflict. In V. J. Derlega & J. Grzelak (Eds.), *Living with other people*. New York: Academic.

Purvanova, R. K. (2014). Face-to-face versus virtual reams: What have we really learned? *The Psychologist-Manager Journal, 17*(1), 2–29.

Rapanta, C., Maina, M., Lotz, N., & Bacchelli, A. (2013). Team design communication patterns in e-learning design and development. *Educational Technology Research and Development, 61*, 581–605.

Rashotte, J., Varpio, L., Day, K., Kuziemsky, C., Parush, A., Elliott-Miller, P., King, J. W., & Roffey, T. (2016). Mapping communication spaces: The development and use of a tool for analyzing the impact of EHRs on interprofessional collaborative practice. *International Journal of Medical Informatics, 93*, 2–13.

Reynolds, R., & Salas, E. (2016). Integrating sports psychology into a human factors framework. *International Journal of Sport Psychology, 47*(3), 224–238.

Salas, E., Rozell, D., Mullen, B., & Driskell, J. E. (1999). The effect of team building on performance: An integration. *Small Group Research, 30*(3), 309–329.

Salas, E., Reyes, D. L., & McDaniel, S. H. (2018). The science of teamwork: Progress, reflections, and the road ahead. *American Psychology, 73*(4), 593–600.

Santos, R., Bakero, L., Franco, P., Alves, C., Fragata, I., & Fragata, J. (2012). Characterization of non-technical skills in paediatric cardiac surgery: Communication patterns. *European Journal of Cardio-Thoracic Surgery, 41*, 1005–1012.

Schecter, A., Pilny, A., Leung, A., Poole, M. S., & Contractor, N. (2017). Step by step: Capturing the dynamics of work team process through relationship event sequences. *Journal of Organizational Behavior, 39*, 1163–1181.

Schneider, K., Klünder, J., Kortum, F., Handke, L., Straube, J., & Kauffeld, S. (2018). Positive affect through interactions in meetings: The role of proactive and supportive statements. *The Journal of Systems and Software, 143*, 59–70.

Seeber, I., Maier, R., & Weber, B. (2013). Macrocognition in collaboration: Analyzing processes of team knowledge building with CoPrA. *Group Decision and Negotiation, 22*, 915–942.

Shannon, C., & Weaver, W. (1949). *The mathematical theory of communication*. Champaign: University of Illinois Press.

Simon, H. A. (1972). Theories of bounded rationality. In C. B. McGuire & R. Radner (Eds.), *Decision and organization* (pp. 161–176). Amsterdam: North-Holland Publishing Company.

Sonnentag, S. (2012). Time in organizational research: Catching up on a long neglected topic in order to improve theory. *Organizational Psychology Review, 2*, 361–368.

Stachowski, A. A., Kaplan, S. A., & Waller, M. J. (2009). The benefits of flexible team interaction during crises. *Journal of Applied Psychology, 94*(6), 1536–1543.

Staveski, S. L., Wu, M., Tesoro, T. M., Roth, S. J., & Cisco, M. J. (2017). Interprofessional team's perception of care delivery after implementation of a pediatric pain and sedation protocol. *Critical Care Nurse, 37*(3), 66–76.

Steele-Johnson, D., Beauregard, R., Hoover, P., & Schmidt, A. (2000). Goal orientation and task demand effects on motivation, affect, and performance. *Journal of Applied Psychology, 85*, 274–284.

Stevens, R., & Galloway, T. (2016). Tracing neurodynamic information flows during teamwork. *Nonlinear Dynamics, Psychology, and Life Sciences, 20*(2), 271–292.

Stout, R. J., Cannon-Bowers, J. A., Salas, E., & Milanovich, D. M. (1999). Planning, shared mental models, and coordinated performance: An empirical link is established. *Human Factors, 41*, 61–71.

Strang, A. J., Funke, G. J., Russell, S. M., Dukes, A. W., & Middendorf, M. S. (2014). Physio-behavioral coupling in a cooperative team task: Contributors and relations. *Journal of Experimental Psychology: Human Perception and Performance, 40*, 145–158.

Straus, S., & McGrath, J. (1994). Does the medium matter? The interaction of task type and technology on group performance and member reactions. *Journal of Applied Psychology, 79*(1), 87–97.

Swigger, K., Hoyt, M., Serçe, F. C., Lopez, V., & Alpaslan, F. N. (2012). The temporal communication behaviors of global software development student teams. *Computers in Human Behavior, 28*, 384–392.

Theureau, J. (2003). Course-of-action analysis and course-of-action centered design. In E. Hollnagel (Ed.), *Handbook of cognitive task design* (pp. 55–81). Mahwah: Lawrence Erlbaum Associates.

Thompson, L., & Coovert, M. (2003). Teamwork online: The effects of computer conferencing on perceived confusion, satisfaction, and post-discussion accuracy. *Group Dynamics: Theory, Research, and Practice, 7*(3), 135–151.

Tiferes, J., & Bisantz, A. M. (2018). The impact of team characteristics and context on team communication: An integrative literature review. *Applied Ergonomics, 68*, 146–159.

Tolston, M. T., Riley, M. A., Mancuso, V., Finomore, V., & Funke, G. J. (2019). Beyond frequency counts: Novel conceptual recurrence analysis metrics to index semantic coordination in team communication. *Behavior Research Methods, 51*, 342–360.

Tschan, F. (1995). Communication enhances small group performance if itconforms to task requirements: The concept of ideal communication cycles. *Basic and Applied Social Psychology, 17*, 371–393.

Tschan, F. (2002). Ideal cycles of communication (or cognitions) in triads, dyads, and individuals. *Small Group Research, 33*(6), 615–643.

Tushman, M., & Romanelli, E. (1985). Organizational revolution: A metamorphosis model of convergence and reorientation. *Research in Organizational Behavior, 7*, 171–222.

Uitdewilligen, S., Waller, M. J., & Pitariu, A. H. (2013). Mental model updating and team adaptation. *Small Group Research, 44*(2), 127–158.

Uitdewilligen, S., Rico, R., & Waller, M. J. (2018). Fluid and stable: Dynamics of team action patterns and adaptive outcomes. *Journal of Organizational Behavior, 39*(9), 1113–1128.

Van den Bulte, C., & Moenaert, R. K. (1998). The effects of RandD team co-location on communication patterns among RandD, marketing, and manufacturing. *Management Science, 44*(11), S1–S18.

van der Haar, S., Koeslag-Kreunen, M., Euwe, E., & Segers, M. (2017). Team leader structuring for team effectiveness and team learning in command-and-control teams. *Small Group Research, 48*(2), 215–248.

van Ruijven, T., Mayer, I., & deBruijne, M. (2015). Multidisciplinary coordination of on-scene command teams in virtual emergency exercises. *International Journal of Critical Infrastructure Protection, 9*, 13–23.

Vasileiadou, E., & Vliegenthart, R. (2014). Studying dynamic social processes with ARIMA modeling. *International Journal of Social Research Methodology, 17*(6), 693–708.

von Cranach, M., Kalbermatten, U., Indermuehle, K., & Gugler, B. (1982). *Goal-directed action.* London: Academic.

von Cranach, M., Ochsenbein, G., & Valach, L. (1986). The group as a self-active system: Outline of a theory of group action. *European Journal of Social Psychology, 16*(3), 193–229.

Wallace, W. A. (1987). Capital costs versus costs-in-use: A content analysis of design team member communication patterns. *Construction Management and Economics, 5*(4), S73–S92.

Waller, M. J. (1999). The timing of adaptive group responses to nonroutine events. *Academy of Management Journal, 42*(2), 127–137.

Wang, L., Doucet, L., Waller, M., Sanders, K., & Phillips, S. (2016). A laughing matter: Patterns of laughter and the effectiveness of working dyads. *Organization Science, 27*(5), 1142–1160.

Wasserman, S., & Faust, K. (1994). *Social network analysis: Methods and applications*. New York: Cambridge University Press.

Webber, C. L., & Zbilut, J. P. (1994). Dynamical assessment of physiological systems and states using recurrence plot strategies. *Journal of Applied Physiology, 76*, 965–973.

Weinberger, A., & Fischer, F. (2006). A framework to analyze argumentative knowledge construction in computer-supported collaborative learning. *Computers and Education, 46*, 71–95.

Weingart, L. (1992). Impact of group goals, task component complexity, effort, and planning on group performance. *Journal of Applied Psychology, 77*, 682–693.

Welp, A., Johnson, A., Nguyen, H., & Perry, L. (2018). The importance of reflecting on practice: How personal professional development activities affect perceived teamwork and performance. *Journal of Clinical Nursing, 27*, 3988–3999.

Wheelen, S., & Hochberger, M. J. (1996). Validation of the group development questionnaire. *Small Group Research, 27*(1), 143–170.

Wittenberg, E., Ferrell, B., Goldsmith, J., Ragan, S. L., & Paice, J. (2016). Assessment of a statewide palliative care team training course: COMFORT communication for palliative care teams. *Journal of Palliative Medicine, 19*(7), 746–752.

Woolley, A. (1998). Effects of intervention content and timing on group task performance. *Journal of Applied Behavioral Science, 34*(1), 30–46.

Wybo, J. (2013). Percolation, temporal coherence of information, and crisis prevention. *Safety Science, 57*, 60–68.

Xiao, Y., Seagull, F. J., Mackenzie, C., Ziegert, J., Klein, K. J. (2003). Team communication patterns as measures of team processes: Exploring the effects of task urgency and shared team experience, Proceedings of the human factors and ergonomics society 47th annual meeting, pp. 1502–1506.

Yoon, S. W., & Johnson, S. D. (2008). Phases and patterns of group development in virtual learning teams. *Educational Technology Research and Development, 56*, 595–618.

Zoethout, H., Wesselink, R., Runhaar, P., & Mulder, M. (2017). Using transactivity to understand emergence of team learning. *Small Group Resaerch, 48*(2), 190–214.

Index

S. McComb, D. Kennedy, *Computational Methods to Examine Team Communication*, Computational Social Sciences,
https://doi.org/10.1007/978-3-030-36159-4

P

Q

R

S

T

V

W

The manufacturer's authorised representative in the EU is Springer Nature Customer Service Centre GmbH, Europaplatz 3, 69115 Heidelberg, Germany. If you have any concerns regarding our products, please contact ProductSafety@springernature.com

Printed and bound by CPI Group (UK) Ltd, Croydon, CR0 4YY
22/09/2026
02221058-0009